KR

Crystalline Symmetries

An Informal Mathematical Introduction

Crystalline Symmetries

An Informal
Mathematical Introduction

Marjorie Senechal

Department of Mathematics
Smith College

Adam Hilger
Bristol, Philadelphia and New York

British Library Cataloguing in Publication Data

Senechal, Marjorie
    Crystalline symmetries : an informal mathematical
    introduction.
    1. Crystallography
    I. Title
    548

ISBN 0-7503-0041-8

Library of Congress Cataloging-in-Publication Data

Senechal, Marjorie.
    Crystalline symmetries : an informal mathematical introduction /
    Marjorie Senechal.
        p.   cm.
    Includes bibliographical references and index.
    ISBN 0-7503-0041-8
    1. Crystallography, Mathematical.   I. Title.
    QD911.S47   1990
    548'.7–dc20   90-41021
                                                                       CIP

The author has attempted to trace the copyright holder of all the figures
and tables reproduced in this publication and apologizes to copyright
holders if permission to publish in this form has not been obtained.

Consultant Editor: Professor R F Streater

Published under the Adam Hilger imprint by IOP Publishing Ltd
Techno House, Redcliffe Way, Bristol BS1 6NX, England
335 East 45th Street, New York, NY 10017-3483, USA

US Editorial Office: 1411 Walnut Street, Philadelphia, PA 19102

Printed in Great Britain by J W Arrowsmith Ltd, Bristol

This book is dedicated to the memory of my father

Abraham Wikler
1910–1981

Contents

Preface

Von Laue's discovery of the diffraction of x-rays by crystals, in 1912, was the key to the solid state revolution of our times, a revolution which has led to the development of powerful techniques for the determination of structure on the molecular and atomic scale. In the ensuing 78 years, crystallography has branched into a large number of subspecialties, and its tools have become central to a large number of fields. Not only mineralogy and crystallography, but also physics, chemistry, biology, materials science, medicine and other branches of science have been fundamentally changed by the availability and power of crystallographic techniques.

Most of the conceptual tools for the classification of crystal structure—the theory of lattices and space groups—had been developed in the nineteenth century, but in the absence of any obvious use for them they were put on the shelf to await their time should it ever come. By 1915 the need for those tools was evident, and so was the need to present them in a form useful to practioners of the new art. Thus was born the series of volumes which became the *International Tables for X-Ray Crystallography*. Over the years, the International Union of Crystallography has revised the *Tables* to keep pace with the needs of workers in a growing number of fields whose scientific backgrounds are less and less coherent. This has meant that not only must more data be supplied, but also more detailed discussions of interpretation of these data are necessary. The most recent edition of these tables is a multi-volume set, of which only volume A, on crystal symmetry, has appeared so far. This volume alone is over 700 pages long.

Unfortunately, despite the extensive discussions of interpretation that they contain, there is an inverse relation between the size of the *Tables* and the ease of deciphering them. Several companion volumes have been written which explain how to use the tables, but there is no volume devoted solely to the simple and elegant mathematical ideas which underly them. As the computer increasingly comes to dominate the practice of crystallography, these simple ideas are in danger of being forgotten, or never even being learned, by the crystallographic community.

For many years I have lamented this gap, especially when asked by mathematicians and scientists to recommend an introductory account of mathematical crystallography. An invitation in 1988 from Dan Shechtman of the Department of Materials Engineering of the Technion, in Haifa, Israel to give a series of six lectures on this subject (including a lesson on how to decipher the *Tables*) gave me an opportunity to try to remedy this situation. At the same time, it was an opportunity to explain some of the ways in which new developments in crystallography, such as the discovery of quasicrystals by Shechtman, are stimulating a reexamination of the mathematical foundations on which the *Tables* rest.

In trying to transform these lectures to the printed word, I have resisted the temptation (never very strong) to write a textbook with proofs and exercises. Detailed, rigorous and sometimes excellent accounts of most of the topics I discuss can be found in the literature; what is missing is an overview. My intent has been to write an account of this overview for the armchair reader, not the reader at a desk with pencil and paper and maybe a copy of the *Tables* too. I doubt that I have succeeded in making it as simple as that, but at least I have avoided giving proofs unless they are especially instructive, and have avoided assigning exercises unless they really look like fun (answers can be found in Appendix 2). Armchair mathematics is like armchair travelling. It is not the real thing, but it can be enjoyable, and it can help you decide if you want to go there to see for yourself. Just as the articles in the New York Times Sunday Travel Section contain a coda entitled 'Getting There' with lists of air service, hotels, and so forth, this book concludes with an annotated bibliography for further study.

The basic ideas of the classical theory are quite simple. The situation is much like trigonometry: despite the plethora of tables, formulas, and graphs, all of trigonometry is really an elaboration of the Pythagorean theorem, and once you appreciate this fact, the main ideas of the subject are very easy to grasp. Nevertheless, as I learned long ago when I tried unsuccessfuly to teach trigonometry to my younger brother in one hour (to save him from failing a calculus course), some mathematical sophistication is necessary in order to comprehend and make use of this simplicity. Thus I will assume that the reader has taken an introductory course in linear algebra. I do not assume prior acquaintance with the idea of a group.

My choice of topics has been guided primarily by the goal of making the *Tables* intelligible in the span of a few lectures. I have also presented some material that is especially beautiful as well as useful but which is not easy to find elsewhere, including Klein's enumeration of the finite rotation groups of the sphere, Delone's derivation of the Voronoï cells of the three-dimensional lattices, and de Bruijn's construction of the Penrose tiles.

Mathematical crystallography has a rich and interesting history that deserves a book of its own. Historical remarks are clustered in several places throughout this book; all the people who appear in them are identified in Appendix 1.

It is a pleasure to thank the faculty and staff of the Department of Materials Engineering of the Technion for their hospitality. I am also indebted to the members of the Mathematics Department of the University of the Philippines at Diliman, Quezon City, for their hospitality and interest in mathematical crystallography; it was there, in 1987, that the need for a monograph like the present one first became clear to me. I am grateful to Richard Roth, Peter Engel, Louis Michel, Ray Streater, Doris Schattschneider and Richard Ghez for their very helpful comments on the manuscript. Thanks also are due to Richard Ghez and Jim Revill for their enthusiasm and encouragement, and to Jen Halford for her skilful editing and her patience.

Marjorie Senechal
August 1990

Chapter 1

Mathematical Crystals

1.1 What is mathematical crystallography?

Many branches of the science of crystals make extensive use of mathematics. For example, mathematical analysis is central to the study of crystal optics and to the interpretation of diffraction patterns. However, for rather obscure historical reasons, *mathematical crystallography* usually means the study of spatial patterns that have properties that make them appropriate models for crystal structure; it has sought to classify both atomic patterns and external forms, to show how they are related to one another, and to predict some of the properties that crystals with these patterns and forms must have.

By 1980, this program seemed to have been achieved. During the nineteenth century, a theory of crystal structure as a modular, repeating pattern had slowly emerged; by 1890 the corresponding catalogue of basic crystalline atomic patterns was complete. Within 25 years, this catalogue, in the form of tables, was widely used in the new field of x-ray crystallography; by the early 1980's, the underlying mathematical ideas had been reformulated and reinterpreted in many different ways and the results stored in large computers. The relation of these patterns to the shapes and properties of ideal crystals was, by that time, also fairly clear. Although many intriguing problems remained unsolved on the theoretical level, it was not anticipated that their solutions would bring with them any major surprises.

But in science there are always surprises. The discovery, in 1984, of crystals which break the symmetry 'rules' of crystallography has prompted a reexamination of the role of crystallography's most treasured theoretical support, the symmetry group. This will proceed whatever structure quasicrystals are eventually determined to have (as of this writing, August 1990, their structure has not yet been completely deciphered). One direction this reexamination is taking is intensive research in the theory of nonperiodic patterns. If this theory continues to develop along useful lines, the mathematical crystallography of tomorrow may be very different from today's.

Before we try to see where the subject is going, however, it is helpful to see what it is now and where it has been. Our approach to history is inspired by the *Guides Michelin*.

1.2 Un peu d'histoire

In his thoughtful and influential monograph 'Order and Life', J Needham pointed out that 'form is simply a short time-slice of a single spatio-temporal entity'. Needham was speaking of biological form, but the remark is equally applicable to the conceptual forms, or paradigms, which guide our scientific thinking. For example, if you open any contemporary textbook on crystallography, you will find a sentence something like this (see Figure 1.1):

The regular shape of crystals suggests that within a crystal atomic building units, congruent to each other, are regularly arranged.

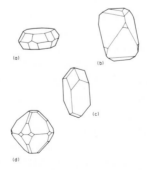

Figure 1.1. What sort of structure do these regular shapes suggest?

However, even a casual study of the history of crystallography shows that the shapes of crystals did not necessarily suggest anything of the sort to our early predecessors. True, Plato had postulated geometric units as the basic forms of the four elements, earth, air, fire and water: to these elements he assigned the cube, the regular octahedron, the regular tetrahedron, and the regular icosahedron (Figure 1.2). These assignments were not entirely arbitrary. In the first place, the octahedron, tetrahedron, and icosahedron are made of identical, equilateral triangles; Plato envisioned that these triangles could be disassociated and regrouped. This would explain, for example, the transition from liquid to steam that water undergoes when it receives a sufficiently large dose of fire. And the assignment of shapes also made some sense: the cube was appropriate for earth particles, since earth sits solidly; the tetrahedron, which has the sharpest corners of the four, was appropriate for fire particles. The octahedron, nice and light, was a good choice for air, while the icosahedron, which is almost spherical, was just the particle for fluids. (Even today it is argued that icosahedral groupings of molecules are fundamental for the structure of liquids and their solid state counterparts, the glasses.)

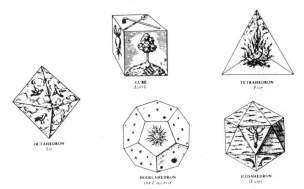

Figure 1.2. Plato assumed that the basic particles of earth, air, fire and water had the forms of regular polyhedra. From Beck, Bleicher, and Crowe, *Excursions into Mathematics*, 1969. Reproduced by permission of Worth Publishers, New York.

Nevertheless, Plato's intriguing hypothesis was persuasively refuted by Aristotle on the grounds that not all of these units could fill space. According to Aristotle, a particulate theory of matter could only be valid if the units of which matter was composed filled all of space without gaps, because vacua do not exist. This argument, together with the apparent inability of any space-filling model to explain motion, contributed to the demise of this early, more or less atomic, theory of matter.

For centuries thereafter, atomic theory lay dormant, for religious as well as scientific reasons. Instead of focusing on shape, subsequent classification schemes for crystals (until the seventeenth century) included 'virtues' such as talismanic and healing powers, and the imagined 'imitative' characteristics of surface markings. The annals of seventeenth century science are replete with debates about the origin of crystals and fossils (which were not always distinguished). Were they 'sports of nature'? Were they permanently frozen ice? Some argued that crystals were of organic origin. The fact that crystals grow, and in some cases have visible 'veins', was seen by many as evidence that crystal structure is analogous to the structure of plants. As late as the eighteenth century, the great botanist C Linnaeus attempted to extend his successful classification scheme for plants to the mineral kingdom. His distinction between mother and father stones strikes our ears as somewhat curious.

The idea that crystals are modular structures—that is, that they can be represented by aggregations of spheres or polyhedral building blocks (the word *polyhedron* means 'many sides'), was proposed early in the seventeenth century by J Kepler, and again some years later by R Hooke (Figure 1.3) and by N Steno.

Their ideas finally came to fruition early in the nineteenth century in the work of R-J Haüy; perhaps not coincidentally, this was also the time of the rebirth of atomic theory. A highly debatable legend has it that this French abbé accidentally dropped a calcite specimen belonging to a friend; to his dismay, the crystal shattered into tiny pieces. But genius turned disaster into triumph: Haüy noted that the tiny pieces all had the same shape, perfect, identical rhombohedra or groupings of these rhombohedra. This suggested to him that crystals are arrays of subvisible blocks whose shapes were specific to the type of crystal (Figure 1.4).

Haüy showed how the same blocks could be used to construct different shapes; in this way he was able to account for the puzzling fact that crystals of the same substance sometimes have different external forms (Figure 1.5).

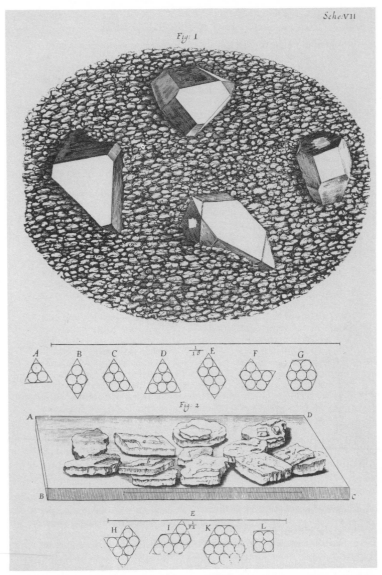

Figure 1.3. 'Had I time and opportunity, I could make probable, that all these regular Figures ... arise onely from three or four several positions of Globular particles.'—Hooke, *Micrographia*, 1665.

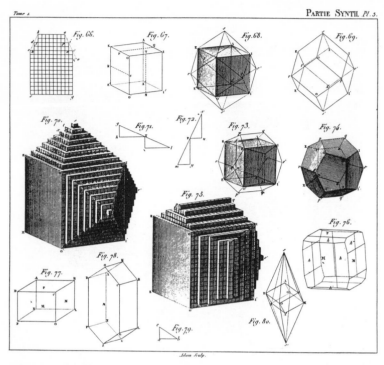

Figure 1.4. Haüy's idea of crystal structure. From *Traité de Crystallographie*, 1822.

A good theory restricts: it must explain why, among all possible phenomena, only a certain few occur. Not every polyhedron occurs as a crystal form; it is necessary to explain why the others do not. Haüy realized that his building-block theory had implications for crystal symmetry as the overall shape of a crystal cannot have symmetry which is impossible for its pattern of constituent parts. Like Plato, though for different reasons, Haüy too believed that it is possible to group the basic particles of a crystal into shapes that fill space. This is a severe restriction. For example, among the regular polygons only triangles, squares, and hexagons fill the plane (Figure 1.6); pentagons and polygons with more than six sides do not. In particular, it follows that a repeating pattern of congruent blocks (e.g., a crystal) cannot have five-fold symmetry.

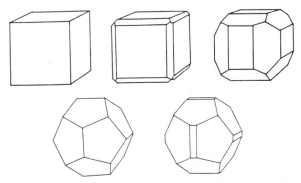

Figure 1.5. Five drawings of pyrite crystals. From V Goldschmidt, *Atlas der Kristallformen*, Carl Winters Universitätsbuchhandlung, Heidelberg, 1913.

(A more rigourous argument will be given later in this chapter.) In three dimensions, the regular icosahedron and dodecahedron have five-fold symmetry; thus Haüy's theory predicts that no crystal can have these shapes. Notice that the dodecahedron in Figure 1.5 is not regular. The 32 classes of symmetry that crystals *can* have were first enumerated in 1826 by M Frankenheim.

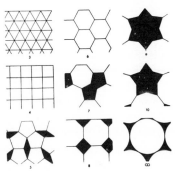

Figure 1.6. Among the regular polygons only triangles, squares, and hexagons fill the plane.

Only after Haüy did it become 'evident' that the regularity of crystal form 'suggests' the regularity of crystal structure. This suggestion generated a great deal of research on 'regular systems of points' (see below) and their symmetries. The symmetry groups of regular systems of points are precisely the space groups, the centennial of whose enumeration we celebrate this year. Eventually this suggestion was elevated to the very definition of 'crystal', and after the discovery of x-ray diffraction the repeating pattern paradigm became the cornerstone of crystal theory.

The success of x-ray diffraction in unlocking the secrets of the solid state, the sharpening of both theory and technique to solve the structures of large biological molecules, and their generalization to wholly new applications such as computer tomography is one of the most remarkable stories of twentieth century science. There is hardly an area of our lives that has not been touched by it, from the watches we wear to diagnostic procedures at the health clinic to the ubiquity of solid state electronics. (Unfortunately, as far as the general public is concerned, it is a story still untold.)

But for all its successes, it may be that the present definition of a crystal is too restrictive, and the focus on regular repetition, or what is the same thing, symmetry, too single-minded. The centrality of symmetry as a fundamental conceptual tool for science is implicit in P Curie's banal but widely quoted 'principle of symmetry':

> *When certain causes produce certain effects, the elements of symmetry in the causes ought to reappear in the effects produced.*

The existence, and the importance, of symmetry, is rarely questioned. Yet we still do not understand where symmetry comes from. Indeed, explaining why atoms should group themselves into regular arrays is a major open problem in statistical mechanics. In fact, symmetry is an effect as well as a cause; to understand what it is the effect of, as well as what its effects are, is a central problem of contemporary mathematical crystallography. When we do understand it, we may find that symmetry is not as axiomatic as once was thought.

1.3   Models of crystal structure

Today, with modern technology at our disposal, we can 'see' the structure of crystals at the atomic level, but we still need models to help us interpret these atomic patterns, and to understand the roles they play in determining other crystal properties.

In our time, all models of crystal structure assume that crystals are built of one or a few different kinds of discrete units, arranged in more or less modular fashion. The models are distinguished from one another by such features as the shapes of these modules and how they are linked to their neighbors to form a potentially infinite structure.

In the three principal models used today, the modules are spheres, space-filling polyhedra, and the nodes of networks. All three models can be formulated in ways consistent with Haüy's theory, but they are more general. It is remarkable that all three can be traced to ideas of J Kepler in the early seventeenth century.

First we need some geometric terminology.

1. An object is *convex* if it has no dips or dents: more precisely, if any two points in or on the surface of the object can be joined by a straight line segment that lies entirely within or on the object. For example, the object on the left in Figure 1.7 is convex; the one on the right is not.

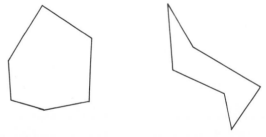

Figure 1.7. The object on the left is convex; that on the right is not.

2. A set of points is *discrete* if the distance between any pair is greater than or equal to some minimum distance. The vertices of a regular polygon constitute a discrete set; the points of a circle do not (Figure 1.8).

3. The word 'graph' is used in two ways in mathematics: in analysis it is a picture of the relation between the domain and the range of a function; in combinatorial theory, a *graph* is a discrete set of points, finite or infinite, together with arcs which connect some or all pairs. In this book all graphs are combinatorial ones; several

Figure 1.8. The vertices of a regular polygon are a discrete set; the points of a circle are not.

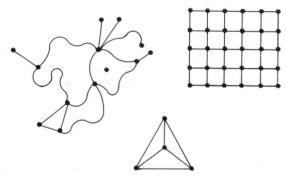

Figure 1.9. Examples of graphs.

examples are shown in Figure 1.9. The points are sometimes called the *nodes* of the graph.

4. A set of points is *relatively dense* in the plane, or in space, if any circle, or sphere, that does not contain any points of the set has a radius less than some fixed number. A relatively dense point set in an unbounded space necessarily contains infinitely many points. (The point sets in Figures 1.16, 1.17 and 1.18 are representative portions of relatively dense sets.)

1.3.1 *Sphere-packings*

In 1611, Kepler wrote an essay on snowflakes as a 'New Year's gift to a friend'. In it, he attempted to explain the hexagonal shapes of snowflakes by assuming that they are made of spherical particles arranged in planar closest packing (Figure 1.10). He reached no definitive conclusions, but the essay contained many geometrical ideas that later proved fruitful in thinking about crystal structure. As we have noted, Kepler's ideas anticipated those of Hooke some fifty years later.

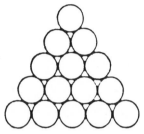

Figure 1.10. Kepler suggested that the internal structure of snowflakes can be modelled by closely packed spheres.

The sphere-packing model continues to be a useful and versatile one. Today, models in which different kinds of atoms are represented by spheres of different sizes or different colors are used routinely in crystallography, especially in crystal chemistry. Sphere-packing models are equally useful in fields remote from crystallography, such as coding theory.

Sphere-packing theory is also a field of active mathematical research: there are many fundamental problems which are still unsolved. The most famous of these is: *What is the densest packing of equal spheres in three-dimensional space?* This problem was part of the 18th of 23 which D Hilbert posed to the international mathematical community in 1900. As C A Rogers has remarked, 'All physicists know and most mathematicians believe' that the densest packings are the well-known cubic or hexagonal closest packings (Figure 1.11). But it still has not been proved that no denser packing exists.

1.3.2 Space-filling polyhedra

In this model, the units are space-filling polyhedra instead of spheres. This model too can be traced to Kepler's New Year's gift. Kepler showed that some space-filling polyhedra can be obtained by uniformly compressing arrays of spheres. He discovered three different space-fillers in this way.

(a)   (b)

Figure 1.11. (a) A second layer of spheres can be nested in the hollows between the spheres of Figure 1.10. There are two choices for the third layer: directly above the spheres of Figure 1.10, or (b) above the unoccupied hollows of (a). Both arrangements are equally dense. It is not known whether a denser packing exists.

As we have seen, the space-filling theory of crystal structure in its more modern form was initiated by Haüy. But neither Kepler nor Haüy seems to have asked the question: *What are the possible shapes of space-filling polyhedra?*

In 1885 the Russian crystallographer E S Fedorov did ask this, and he answered it in part. Fedorov showed that there are exactly five types of convex polyhedra that can fill space when juxtaposed face to face in parallel arrays (Figure 1.12).

The more general problem of finding all possible space-filling polyhedra is exceedingly difficult; indeed it may be hopeless. Support for despair is derived from the fact that no algorithm exists for deciding whether a given two-dimensional shape will tile the plane.

Tiling theory in three-dimensional space is subtle. For example, consider tetrahedra, polyhedra with four triangular faces. Some kinds of tetrahedra fill space, but others do not. We still do not have a complete answer to the question: *Which tetrahedra fill space?* As Figure 1.13 shows, copies of a regular tetrahedron do not fill space—although, curiously, Aristotle thought that it did!

Nor do we know the maximum number of faces that a space-filling polyhedron can have. So far, the maximum number of faces is 38 (Figure 1.14); this was found by P Engel with a computer

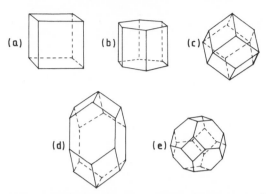

Figure 1.12. Fedorov's five space-filling polyhedra: (a) cube, (b) hexagonal prism, (c) rhombic dodecahedron, (d) elongated dodecahedron, (e) truncated octahedron.

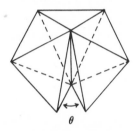

Figure 1.13. Aristotle thought that the regular tetrahedron fills space; had he constructed simple models, he would have seen his error.

search in 1980.

1.3.3 Networks

In this model the units are represented by points joined by line segments (not necessarily straight): in other words, the model is an infinite graph. The graphs of interest in crystallography have regularity properties, such as combinatorially identical nodes. In two dimensions, the networks in which all nodes have this property are sometimes called Kepler graphs. Kepler worked out the 11 possible infinite tilings of the plane with regular polygons and congruent configurations at each vertex in 1619; they are shown, together with some partial tilings which cannot be extended indefinitely, in Figure 1.15—which are which? In fact, if we relax the

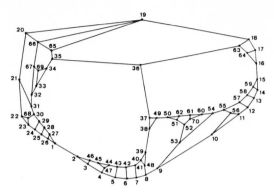

Figure 1.14. A schematic drawing of a space-filling polyhedron with 38 faces.

requirements of regularity and even congruence, it still turns out that these are the only planar networks with combinatorially identical nodes. This means, in particular, that there are no infinite planar graphs whose cells all have more than six sides.

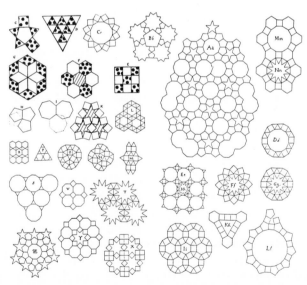

Figure 1.15. Kepler explored tiling the plane with regular polygons. He found all the infinite tilings with identical nodes, and others that cannot be extended forever.

Kepler's investigations also led him to explore nonperiodic patterns. Notice in particular the configuration of tiles labelled Aa, in which he tried to arrange regular pentagons as symmetrically as possible. We will look at this configuration more closely in Chapter 7.

Three-dimensional networks have been investigated by crystallographers for many years, and several important classes have been studied in detail. The complete list of networks with combinatorially identical nodes in three-dimensional space is still not known; there is no comprehensive theory for three-dimensional networks as there is in the planar case, because the methods used to solve the problem in the plane cannot be extended to higher dimensions.

Sphere-packings, space-filling polyhedra, and infinite graphs are all useful models of crystal structure. Which model one wants to use depends on the context. From an abstract point of view, the three are interchangeable (though not identical).

1.3.4 (r,R) systems

The three models we have just discussed can all be interpreted as patterns of points: the centers of the spheres, the centroids of the polyhedra, the nodes of the graphs. By considering only the points, and not the spheres or polyhedra surrounding them or the arcs joining them, we can create a more abstract model for crystal structure. We lose the particular features of each model that makes them valuable in the study of real crystals, but we gain a deeper understanding of some of the properties which underly them all.

Let us be very abstract. One very general kind of point set is the type known as an (r, R) *system*. These sets, introduced by B N Delone, have two key properties: they are discrete (with minimum distance r), and they are relatively dense (with maximum 'empty hole' radius R). A portion of an (r, R) system is shown in Figure 1.16.

Obviously, since they satisfy so few axioms, (r, R) systems are very general. The disordered set in Figure 1.16 might be a model for a monoatomic gas; the highly ordered point set in Figure 1.17 is also an (r, R) system.

In classical crystallography, the (r, R) systems of interest are very orderly. For example, in the simplest sphere-packing and space-filling models, the points are usually arranged in identically

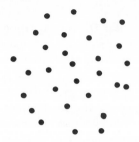

Figure 1.16. A portion of an (r, R) system.

spaced, parallel rows and planes. Such an array is called a *lattice*. (The word is deliberately intended to suggest the lattices familiar to us in fences and grids; point lattices share many properties with them.)

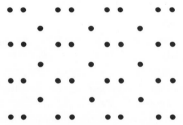

Figure 1.17. A highly ordered (r, R) system.

The points of a lattice are related by shifts called *translations*; they are indicated by arrows in Figure 1.18. A pattern whose symmetry includes translations is said to be *periodic*.

More generally, when the points of an (r, R) system are arranged so that the configuration looks the same from any point of it, the system is said to be *regular*. (The lattice is just the simplest example.) By its very definition, a regular system of points has a very special property: its points are equivalent. That is, if we had two identical copies of the system, we could superimpose any point of one on any point of the other, and at the same time superimpose all the points of the two systems.

Does regularity imply periodicity? The answer is yes: about one hundred years ago, A Schoenflies proved that

Figure 1.18. A plane lattice.

Theorem 1.1. *A regular (r, R) system in three-dimensional space has three independent translation directions, i.e., it is periodic.*

(Later L Bieberbach showed that the analogous result is true in space of any dimension.)

This means that a regular point set is either a single lattice or the union of a finite number of translates of a single lattice (look for the lattices in Figure 1.19).

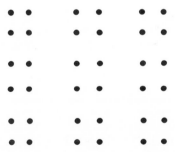

Figure 1.19. This regular (r, R) system consists of four congruent lattices.

Using lattices, it is easy to discover the restriction on crystal symmetry that Haüy found in 1822. This result is known today as the *crystallographic restriction*.

Theorem 1.2. *Five-fold rotational symmetry is incompatible with a two- or three-dimensional translation lattice; so are seven-fold, eight-fold, and all higher rotational symmetries.*

Figure 1.20 shows why. Since a lattice is a regular (r, R) system, there is a minimum distance r between its points and all points are equivalent. Suppose two points P and Q lie on axes of five-fold rotation, and that the distance from P to Q is the minimum distance r. As we rotate about P successively through angles of 72°, Q is copied five times; if we then rotate about Q, P is copied five times. Elementary geometry shows that two of the points generated in this way (Q' and P' in Figure 1.20) are closer than r, which is impossible according to our hypothesis.

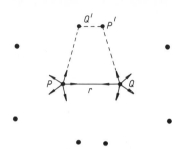

Figure 1.20. A lattice cannot contain centers of five-fold symmetry (see text).

It follows from Theorems 1.1 and 1.2 that the number of symmetry types of regular (r, R) systems is finite. This allows us to classify and enumerate them, and thus to classify mathematical crystals.

All of the models we have discussed presuppose the existence of an infinite pattern. It is important to stress, however, that we are still trying to understand why the infinite patterns we call crystals usually turn out to be regular. How can the juxtaposition of modules generate regularity? Although the symmetry of real crystals is not due only to their geometry and algebra, it is a remarkable fact that geometry and algebra may have something to say about this. We can construct regular (r, R) systems by imposing order only on *finite* regions of otherwise unstructured point sets. More precisely, there is a number M_n, which depends only on the dimension n of the set, such that

Theorem 1.3 (Delone). *If the configurations of points about each point of an (r,R) system which lie within a sphere of radius M_n are congruent, then the system is regular.*

Thus local geometric order in a sufficiently large but finite area forces global or overall regularity.

Is there a direct progression of order properties from the random gas state to the crystalline state? Can we characterize nonperiodic but still orderly patterns of points by weakening the requirement of local order (for example, decreasing the radius M_n or loosening the requirement of congruence)? The study of this problem is part of the reexamination of the foundations of mathematical crystallography.

Chapter 2

Symmetry and Point Groups

2.1 Introduction

Everyone agrees: the objects in Figure 2.1 are symmetrical. But what do we mean by symmetry?

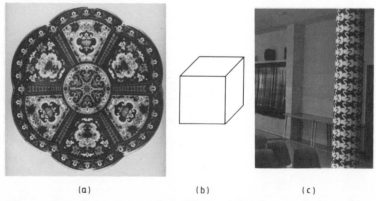

(a)   (b)   (c)

Figure 2.1. Some symmetrical objects: (a) a decorated plate, (b) a cube, (c) a tiled pillar. The plate can be found at the Allentown Art Museum, Allentown, Pennsylvania; the pillar, designed by M C Escher, is in the Johanna Westerman School in The Hague (photograph by J Dijkstra).

If you look up the word 'symmetry' in a dictionary, you will find a definition such as 'due arrangements or balancing of parts with respect to a whole; congruence, harmony'. This does indeed give some idea of what symmetry means, but we need a definition which is more precise. What do we mean by 'parts', and what do we mean by 'arrangement'?

Let us look at the objects in Figure 2.1 a bit more closely with these questions in mind. We see that, in (a), the decorator has conveniently divided the whole into parts, and has emphasized the division by coloring the parts differently. Ignoring the colors (until Chapter 5), we see that the plate is divided into six identical sectors.

The sectors are more than identical: they are *equivalent*, in the sense that they are all arranged in the same way with respect to the whole plate: if we rotate the plate around its center point by multiples of 60°, any of the sectors can be brought to the position originally occupied by any other. We can also interchange the positions of the sectors by reflecting the pattern across the diameters that lie along the boundaries between the sectors, or the diameters that bisect them.

Figure 2.2. Half sectors are equivalent parts of the pattern on this plate.

Notice that these sectors themselves are divided into equivalent parts by their bisectors (Figure 2.2). The process stops here: any further subdivision results in parts which are not even congruent to one another.

We turn next to the cube. It has six equivalent square faces, each composed of eight equivalent sectors. This means that, if we dissect the cube as far as possible, we end up with 48 equivalent parts (Figure 2.3).

The swans swimming around the pillar (Figure 2.4) are all equivalent, if again we ignore differences in color. They have no symmetry of their own, so they cannot be subdivided.

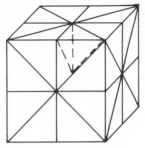

Figure 2.3. There are many ways to divide a cube into equivalent parts; the maximum number is 48.

Figure 2.4. These swans are all equivalent.

Our rather vague notions of balance and arrangement depend on the existence of equivalent parts in a pattern. As we have seen, the 'parts' into which a pattern can be divided are equivalent if there are motions which carry or 'map' any part to the original position of any other, while leaving the appearance of the object unchanged. Such motions are called *symmetry operations* for the object. Symmetry operations are rigid motions, motions which do not alter the distances between points; rigid motions are also called *isometries*. We will discuss isometries in detail in the next section; for now, note that: (i) the symmetry operations for the plate in Figure 2.1 include rotation about its center point through multiples of $60°$, and reflection across any of the six diameters which divide it into two identical halves; (ii) there are 48 different motions which bring the cube into self-coincidence: some are rotations, some are reflections, and there are also compound operations which involve both rotation and reflection; (iii) if the pillar were infinitely long, it could be brought into self-coincidence by translations. Translations can be repeated *ad infinitum*; thus the infinite pillar has infinitely many distinct symmetry operations.

A symmetry operation may keep some points fixed in place; the set of such points is called the *symmetry element* for the operation. For example, in three-dimensional space the symmetry element of a rotation is a line (the rotation axis) and the symmetry element of a reflection is a plane.

2.2 Symmetries and isometries

Just as symmetry operations are associated with symmetry elements, isometries are classified by the sets of points that they keep fixed. This enables us to classify all isometries.

In three-dimensional space, there are seven kinds of isometries.

If an isometry fixes *four* or more noncoplanar points, then it fixes every point of space. This is the 'trivial' or '*identity*' isometry.

If an isometry fixes at most *three* noncollinear points, then it fixes the whole plane in which these points lie, while all of the points not on this plane are moved to positions at the same distance from the plane as their original position. This is only possible if the motion is a *reflection* across that plane (Figure 2.5).

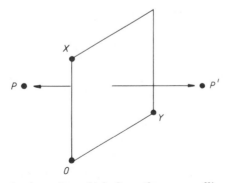

Figure 2.5. An isometry which fixes three noncollinear points is a reflection: plane *XOY* is fixed and *P* and *P'* are interchanged.

If an isometry fixes *two* points, it fixes every point on the line joining them, because it preserves the distances between all pairs of points. The same reasoning shows that if all the fixed points of an isometry lie on a single line, then the motion must be *rotation* about that line (Figure 2.6)

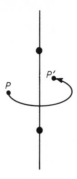

Figure 2.6. An isometry which fixes a single line is a rotation.

Are there isometries which fix exactly *one* point? The answer is yes. These isometries are rather subtle, and for this reason they were discovered somewhat later than the others. They are *rotary reflections*, rotation followed by reflection in a plane orthogonal to the rotation axis, or vice versa (Figure 2.7).

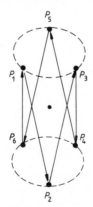

Figure 2.7. Rotary reflection, which combines rotation and reflection, has a single fixed point.

When the angle of rotation is 180°, the isometry is usually called *inversion* because it interchanges all vectors x and $-x$. The fixed point is called a *center of inversion*, or a *center of symmetry*. Figures with this symmetry are said to be *centrosymmetric* (Figure 2.8).

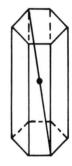 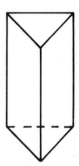

Figure 2.8. A regular hexagonal prism is centrosymmetric; a triangular prism is not.

In the literature, one often finds 'rotary inversion', instead of rotary reflection. Rotary inversion is the composition of rotation about an axis with inversion through a point on that axis. The choice between rotary reflection and rotary inversion is a matter of taste; they do not coincide completely but the same list of isometries can be derived with either one.

The remaining isometries fix *no* points at all. One of them is *translation*, which is just a shift of the whole pattern by some vector amount. The other two combine translation with reflection or with rotation (Figure 2.9). They are known as *glide reflection* and *screw rotation*, respectively.

The three isometries which involve reflection reverse orientations; the four other isometries (including the identity) preserve them. An isometry involving translation can be a symmetry operation only for objects which are infinite in extent.

Artifacts and ornamental art are wonderful sources of symmetry. It is instructive to look for the symmetry elements and symmetry operations for each of the objects in Figure 2.10.

A little linear algebra goes a long way in symmetry theory, so

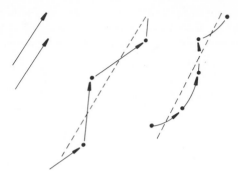

Figure 2.9. Isometries which involve translation have no fixed points: (a) translation, (b) glide reflection, (c) screw rotation.

Figure 2.10. What symmetry operations do these objects have?

let's use it. Isometries which do not involve translation are orthogonal linear transformations. By definition, all of the eigenvalues of an isometry, whether real or complex, must have modulus 1 because an isometry cannot change any distances. It follows that, in three dimensions, for appropriate choices of basis the matrices describing the symmetry operations of a figure can be written either in the form

$$\begin{pmatrix} \pm1 & 0 & 0 \\ 0 & \pm1 & 0 \\ 0 & 0 & \pm1 \end{pmatrix} \quad \text{or} \quad \begin{pmatrix} \cos\theta & -\sin\theta & 0 \\ \sin\theta & \cos\theta & 0 \\ 0 & 0 & \pm1 \end{pmatrix}.$$

The matrix on the right represents a rotation (+1) or a rotary reflection (−1). The diagonal matrices are the identity isometry, a reflection, a 180° rotation, or a 180° rotary reflection (inversion), depending on whether the number of −1's on the diagonal is 0, 1, 2, or 3.

The matrix formulation is helpful in classifying isometries in higher dimensions; we will return to this in Chapter 7.

2.3 Symmetry groups

Some of the basic concepts of symmetry are presented in this section; others will be introduced when we need them, in Chapter 4.

2.3.1 Definitions and axioms

The set of symmetry operations of a object consitutes its *symmetry group*; we will usually denote symmetry groups by the letter G. The symmetry operations belonging to G are the 'elements' of G; do not confuse this use of the word element with 'symmetry element'!

Groups are sets whose elements g_1, g_2, ... can be 'multiplied'. This multiplication may be very different from ordinary number multiplication, but like number multiplication it satisfies the following rules:

(i) G is closed with respect to multiplication: if g_1 and g_2 are in G, so is g_1g_2;

(ii) multiplication is associative: $(g_1g_2)g_3 = g_1(g_2g_3)$;

(iii) G contains an identity element e which plays the same role as 1 in number multiplication and 0 in number addition: $ge = eg = g$ for every element g in G;

(iv) for every g in G there is an element g' in G such that $gg' = g'g = e$. The element g' is the *inverse* of g and is usually denoted by g^{-1}.

Notice that multiplication need not be commutative: it is not required that $g_1g_2 = g_2g_1$. (When multiplication is commutative, it is customary to use additive notation: $g_1 + g_2 = g_2 + g_1$.)

If every element of G can be written as a product of elements of some subset $\{g_1, \ldots, g_k\}$, then this subset *generates* G. The choice of generators is not unique. A group with a single generator is said to be *cyclic*.

For symmetry groups, multiplication should be interpreted to mean 'followed by'. For example, if g is a rotation through 90° and h is reflection in a plane containing g's axis then hg, the rotation g followed by the reflection h, is a reflection in another plane containing g's axis. If $h = g$, then gg, or g^2, is a rotation through 90° followed by another rotation through 90° about the same axis; that is, it is a rotation through 180°. Since $g^4 = e$, this g generates a cyclic group of four elements, g, g^2, g^3, and e. It is worthwhile to take a moment now to verify that axioms (ii) and (iv) are satisfied by the symmetry groups of the objects in Figures 2.1 and 2.10.

The total number of elements of a group G, always including the identity element e, is the *order* of G, which we denote by $|G|$. In Figure 2.1, the orders of the symmetry groups of the objects are 12, 48, and ∞, respectively.

2.3.2   Orbits

To see how a group works, we study its action by following the path of a point as the symmetries in G are performed in succession. For example, if G is the group generated by a rotation g through 60°, and x is a point not lying on the axis of rotation, then x, gx, g^2x, ..., g^5x are six coplanar points that form the vertices of a regular hexagon (Figure 2.11). Such a set is called an *orbit* of G; 'orbit' is the technical name for a set of points which are equivalent with respect to a given group.

If x had been chosen to lie on the rotation axis, then the orbit of x with respect to this same group G would consist of the single point $\{x\}$. Whenever G is a finite group, its orbits are finite sets of points. In general, a finite group has orbits of several different sizes. Figure 2.12 shows six very different orbits for a cube.

By its very definition, a regular (r, R) system is an orbit; in this case the group is infinite.

2.3.3   Subgroups and stabilizers

Sometimes a subset of the elements of the group itself satisfies the group axioms (this will always be the case if the subset is closed and contains the inverse of its elements). Such a subset is called

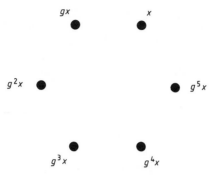

Figure 2.11. The points $x, gx, \ldots, g^5x$ are an orbit for a cyclic group of order six.

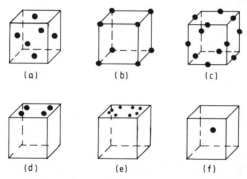

Figure 2.12. Six different orbits for the symmetry group of a cube. (For clarity, (d) and (e) only show points on one (instead of all six) face(s).)

a *subgroup*. For example, the symmetry operations for the plate in Figure 2.1(a) which map dark sectors to dark sectors and light sectors to light constitute a subgroup of the symmetry group of the plate. Strictly speaking, $\{e\}$ and G itself are also subgroups of G; the subgroups intermediate between these two are called *proper* subgroups.

One important kind of subgroup is the *stabilizer* of a point of an orbit. If a point x of an object is left fixed by some of the operations in its symmetry group, then those operations which fix it form a subgroup $S(x)$, the stabilizer (subgroup) of x. In physics, these subgroups are called *isotropy groups* or *little groups*; in crystallography, they are called *site-symmetry groups*. Figure 2.13 shows the symmetry elements of the stabilizers of points of orbits (a)–(d) in Figure 2.12. The stabilizers of points of orbit (e) contain only the identity; the stabilizer of the center of the cube (f) is the full symmetry group of the cube.

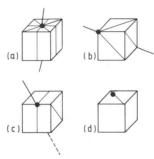

Figure 2.13. The symmetry elements of the stabilizers of the points of orbits (a)–(d) in Figure 2.12.

The stabilizers of the points in the *same* orbit are closely related:

Theorem 2.1. *If $y = gx$ then $S(y) = gS(x)g^{-1}$.*

Subgroups related to each other as H and gHg^{-1} are *conjugate*. Conjugate subgroups are *isomorphic*, which means that there is a one to one correspondence between their elements that is preserved when corresponding elements of each group are multiplied. Conjugate subgroups can be thought of as the same sets of symmetry operations acting in different locations. Thus if we consider conjugate subgroups to be equivalent (and, technically, they are), we can speak of 'the' stabilizer of the points of an orbit.

Notice that the orders of the stabilizer subgroups of the six orbits of the cube, shown in Figure 2.12, are respectively 6, 8, 4, 2, 1 and 48, depending on whether the points of the orbit lie on the cube's face centers, vertices, at an intersection of two mirror planes, on one mirror plane, in a general position, or at the center of the cube. Correspondingly, the number of points in these orbits is 8, 6, 12, 24, 48 and 1.

In general, if G is finite, if s is the order of the stabilizer of an orbit of G, and if n is the number of points in that orbit, then

Theorem 2.2. $|G| = sn$.

An important special case is when $s = 1$. Then each point of the orbit corresponds to a unique operation of G. Such an orbit is said to be *generic* (Figure 2.14). The one to one correspondence between elements of G and points of a generic orbit gives us considerable insight into the action of a group. That is why, for example, we often subdivide tilings and other patterns into their smallest equivalent parts; these parts can be identified with a generic orbit for the symmetry group.

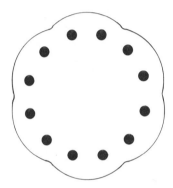

Figure 2.14. A generic orbit for the symmetry group of the plate.

Theorem 2.2 implies that the order s of a stabilizer subgroup is a divisor of $|G|$ (though not every divisor of $|G|$ appears as the order of a stabilizer). (This is a special case of a more general result, *Lagrange's Theorem*, which we will prove in Chapter 4: the order of *any* subgroup H of a finite group G divides $|G|$.) The number $|G|/|H|$ is the *index* of H in G.

2.4 Point groups in 3-space

The symmetry of a crystal is both 'macro' and 'micro': on the one hand there is the symmetry of the 'ideal' crystal form and on the other the symmetry of the crystal's atomic pattern. The theory of space groups (Chapter 4) is the mathematical construction which relates them.

In this section, we will study finite groups of isometries acting on all of three-dimensional space; we do not assume the presence of a lattice. The finite groups associated with crystals belong to this family of groups, and we will see why there are 32 of them.

Finite groups of isometries are also known as *point groups*, for the following reason.

Theorem 2.3. *At least one point is fixed by all of the operations of any finite group of isometries.*

The set of fixed points of the group always includes the centroid of the points of (any of) its orbits. Since isometries preserve the distances between points, all the points of an orbit of a point group are located at the same distance from the centroid. Thus they lie on the surface of a sphere (Figure 2.15). By projecting all the orbits from their common center, we can represent all of them on the same sphere, when that is convenient for us.

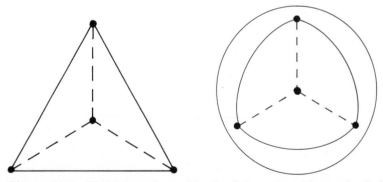

Figure 2.15. The points of an orbit of a finite group of isometries lie on the surface of a sphere.

The group of all symmetries of a sphere is called $O(3)$, which stands for 'the orthogonal group in three-dimensional space'. (The word orthogonal comes from the fact that the matrix representation of an element of this group can be written so that its rows are orthogonal to its columns.) $O(3)$ is not a finite group, since it includes rotations through all angles about any axis through the sphere. We are interested in its finite subgroups. The list of these subgroups is small, and their derivation is a useful exercise in the group-theoretic concepts we have been discussing. We need two theorems:

Theorem 2.4. *(L Euler, 1750) The product of two rotations of a sphere is also a rotation of that sphere.*

This tells us that, if we include the identity operation e, the set of rotations of the sphere is closed. It follows that the rotations belonging to a finite symmetry group G of the sphere form a subgroup H. The index of this subgroup is always two:

Theorem 2.5. *If G is a finite subgroup of $O(3)$ with rotation subgroup H, then either $G = H$ or $|G| = 2|H|$.*

To derive the finite subgroups of $O(3)$, we first present the simple but elegant enumeration of the finite groups of *rotations* given by F Klein about one hundred years ago. Then we will show how it can be extended to include all finite groups of isometries. Finally, we apply the 'crystallographic restriction' (Theorem 1.2) to obtain the list of point groups of crystals.

We let H be a finite group of rotations. To each element $r \neq e$ of H there corresponds a rotation axis which intersects the sphere in two points; these points are called the *poles* of the rotation. We mark the poles on the surface of the sphere (Figure 2.16).

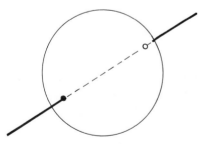

Figure 2.16. The poles of a rotation are points on the surface of a sphere.

The poles of the different axes of the rotations in H may belong to several different orbits of H, say k different ones. We can

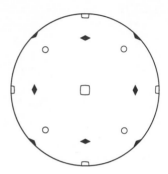

Figure 2.17. The poles of H belong to different orbits.

distinguish the orbits by different colors or symbols (Figure 2.17). For each orbit, $i = 1, 2, \ldots, k$ we have $n_i s_i = |H|$.

We now count the pairs (h, x), where h, a rotation, is in H and x is a pole of h. We do this in two different ways. First, since there are $|H|$ elements of H, the $|H| - 1$ elements not equal to e have two poles each. Thus the total number of pairs is

$$2(|H| - 1). \tag{2.1}$$

On the other hand, there are k orbits of poles, each containing $n_i = |H|/s_i$ points, $i = 1, \ldots, k$, and each point is fixed by $s_i - 1$ rotations. Thus the total number of pairs is

$$\sum_{i=1}^{k} n_i(s_i - 1)$$

which is equal to

$$|H| \sum_{i=1}^{k} \left(1 - \frac{1}{s_i}\right)$$

and to

$$|H| \left(k - \sum_{i=1}^{k} \frac{1}{s_i}\right). \tag{2.2}$$

Equating (2.1) and (2.2), we find that s_i, $|H|$ and k must satisfy the condition

Theorem 2.6. $1/s_1 + \cdots + 1/s_k = k - 2 + 2/|H|$.

The entire list of finite groups of symmetries of the sphere can be extracted from this equation. For example, we can deduce that k, the number of different orbits of poles, must be less than 4. For, since each s_i is the order of the stabilizer of a pole, it must be at least 2. This means that the left-hand side of the equation is at most $k/2$. Thus

$$k/2 \geq k - 2 + \frac{2}{|H|},$$

so

$$k/2 < 2 \quad \text{and} \quad k < 4.$$

Also, if $k = 1$, then $1/s_1 = -1 + 2/|H|$; this is only possible if $|H| = 1$, which means that H contains only the identity. Leaving this trivial case aside, we find that the only cases we need to consider are $k = 2$ and $k = 3$.

The case $k = 2$. The equation of Theorem 2.6 becomes

$$\frac{1}{s_1} + \frac{1}{s_2} = \frac{2}{|H|}.$$

The only solution is $s_1 = s_2 = |H|$, which means that $n_1 = n_2 = 1$. In other words, there are two orbits of one pole each. Since each axis has two poles, H consists of the rotations about a single axis— it is a cyclic group. The poles belong to different orbits because there are no other rotations in H to interchange them. $|H|$ can be any positive integer m which is greater than 1; it is thus the group of rotations of an m-gonal pyramid (Figure 2.18).

The case $k = 3$. The equation now is

$$\frac{1}{s_1} + \frac{1}{s_2} + \frac{1}{s_3} = 1 + \frac{2}{|H|}.$$

There are only four solutions in integers:

$$\frac{1}{2} + \frac{1}{2} + \frac{1}{m} = 1 + \frac{2}{2m},$$

where m is any positive integer, and

$$\frac{1}{2} + \frac{1}{3} + \frac{1}{3} = 1 + \frac{2}{12},$$

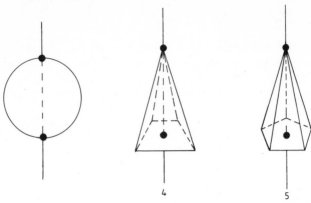

Figure 2.18. Polyhedra whose rotation groups correspond to the case $k = 2$ include the pyramids.

$$\frac{1}{2} + \frac{1}{3} + \frac{1}{4} = \frac{2}{24}$$

and

$$\frac{1}{2} + \frac{1}{3} + \frac{1}{5} = 1 + \frac{2}{60}.$$

These solutions force the relative positions of the poles on the sphere, because every symmetry of H must preserve each orbit.

For example, in the first case above the stabilizer subgroups of two of the orbits are of order two, so the corresponding rotations must be two-fold. There are $|H|/2$ poles in each of these orbits. The third orbit has just two poles, say P and Q, and each has a stabilizer of size $|H|/2$. This means that P and Q must be poles of the same rotation axis, and rotations about the axes corresponding to the poles in the other orbits must interchange them. This is only possible if the other poles lie on the equatorial circle equidistant from P and Q. Moreover, the poles of the other two orbits must alternate on the circle, and must be equally spaced, if they are to correspond to two-fold rotations which map each orbit onto itself (Figure 2.19). H is thus the group of rotations of an m-gonal prism.

The three remaining solutions correspond to the rotations of the tetrahedron, the cube (and the octahedron) and the icosahedron (and pentagonal dodecahedron). Polyhedra with these rotation groups are shown in Figure 2.20.

We can complete the list of point groups by adding reflections and rotary reflections to H. Since, when G contains such opera-

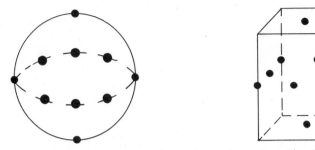

Figure 2.19. The solution $1/2 + 1/2 + 1/m = 1 + 1/m$ forces this arrangement of poles.

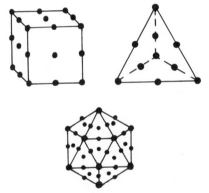

Figure 2.20. The three other solutions correspond to the rotation groups of the regular solids.

tions, we have $|G| = 2|H|$, we just need to add one new operation to H: closure ensures that we will get $|H|$ new operations altogether by multiplying the one we add by all the elements of H. The key to enumeration is the fact that, since H contains all the rotations in G, this new operation must map the configuration of poles onto itself: it cannot introduce any new ones. Because for each i, $s_i n_i = |H|$, and since $|G| = 2|H|$, G doubles either the size of the stabilizer of, or the number of points in, each of the orbits of H. In the first case it preserves the orbits of poles; in the second case it interchanges them (this is of course only possible if H has two orbits of the same size).

For example, if we add a reflection g to the group of rotations of the m-gonal prism, the corresponding reflection plane must either coincide with the equatorial plane, or must be orthogonal to it and contain the m-fold rotation axis (Figure 2.21). In the second case, the reflection can either map each of the two equatorial orbits onto themselves, or it may interchange them. There are no other possibilities.

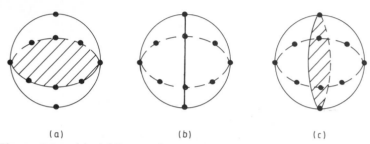

(a)   (b)   (c)

Figure 2.21. (a): Adding a reflection plane orthogonal to an m-fold rotation axis; (b) and (c): adding a reflection plane containing the rotation axis.

The finite subgroups of $O(3)$ are listed in Appendix 3.

Because the crystallographic restriction (Theorem 1.2) forbids rotational symmetry of any order other than 2, 3, 4 or 6, only a few of the groups on our list can be the symmetry group of a crystal with a periodic structure. Eliminating those which do not obey the crystallographic rules, we find a total of 32 groups which *are* possible for crystals. These are the famous crystallographic point groups or crystal classes (see Chapter 6), which have been used in crystallography since the early nineteenth century. But we should not forget the forbidden groups—we will need them in the future.

Chapter 3

Lattices

3.1 Lattices and symmetry

The concept of a lattice was first introduced around 1825, as a way of getting around a conceptual problem posed by Haüy's building blocks. Haüy's critics were puzzled by the physical interpretation of the blocks: what exactly were they supposed to represent, and why should they have the particular shapes that Haüy assigned to them? Their solution was to avoid the question by replacing the blocks by abstract points, which could stand for any physical entity that the reader wanted them to. Like the blocks, these points were equally spaced in parallel rows, the rows equally spaced in parallel planes. This set of points is the configuration we call a lattice.

The lattice has been the framework of crystallography for 160 years, and a magnificent edifice has been built on it. It is not always the most important feature of the structure of a real crystal (for example, there may be thousands of atoms in each repeat unit, and the configuration of these atoms may be of much greater interest than the background lattice); indeed, lattices are incompatible with the observed features of some diffraction patterns, such as those produced by quasicrystals. But lattices are central to classical crystallographic theory.

A three-dimensional lattice is an orbit of a symmetry group, a group generated by three linearly independent translations. We will denote these translations by vectors, say a, b, and c. The points of the lattice are the endpoints of the linear combinations $ua + vb + wc$, where u, v, and w are integers. More generally, the *dimension* of a lattice is the number of linearly independent

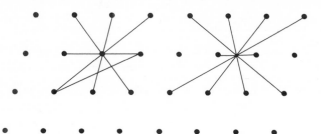

Figure 3.1. Every lattice point, and the midpoint between any pair of lattice points, is a center of symmetry.

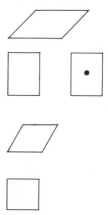

Figure 3.2. There are four crystal systems in the plane, and five Bravais lattices.

translations needed to generate it. The general notation that we will use for lattices is Λ.

The translations form a proper subgroup of the symmetry group of a lattice; the symmetry group also includes the stabilizers of the lattice points (since all lattice points are equivalent, all their stabilizers are conjugate). The stabilizer of a lattice point always includes inversion through that point (Figure 3.1). That is, *every lattice point is a center of symmetry* for the lattice as a whole. Notice that the *midpoint* between any two lattice points is also a center of symmetry for the lattice.

Lattices are assigned to *crystal systems* according to their stabilizer groups: there are four systems in two dimensions, seven in three (Figures 3.2 and 3.3). Each three-dimensional crystal system is a class of conjugate finite subgroups of $O(3)$. (Another widely used classification scheme defines *six* systems; we will compare the two approaches in Chapter 6.)

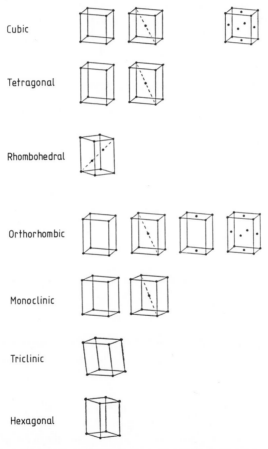

Figure 3.3. The 14 Bravais lattices are distributed among seven crystal systems.

As Figures 3.2 and 3.3 show, very different lattices can belong to the same crystal system. Thus there are five types of lattices in the plane, and 14 lattice types in three-dimensional space. The latter were first enumerated correctly by A Bravais (about 1850), and so they are known as Bravais lattices.

3.2   Unit cells

The generators of a three-dimensional lattice form three edges of a box with the origin as a common vertex. If we complete the box we have a *unit cell*: a parallelepiped with a lattice point at each vertex and no lattice point anywhere else inside it or on its surface (Figure 3.4).

The entire lattice can be reconstructed from a single unit cell by stacking these boxes together. Since each lattice point is shared by eight boxes, and each box has eight vertices, this works out to one lattice point per unit cell.

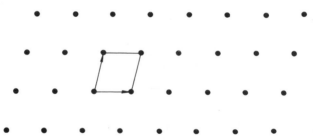

Figure 3.4. A unit cell for the lattice of Figure 3.1.

Any three independent lattice vectors whose edges form such a box can generate the lattice; the choice of *a*, *b*, and *c* is not unique. Thus the unit cell of a given lattice can be chosen in infinitely many different ways, but all of them have the same volume. This volume is of course the product of the 'height' of the cell and the area of its 'base'. Any two of the three generating vectors define a base, and generate a two-dimensional lattice with this base as the unit cell. The *density* of the two-dimensional lattice is the area of the cell. The third vector translates copies of this lattice to form the

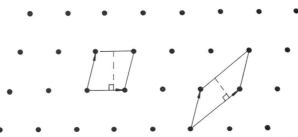

Figure 3.5. Each pair of generators defines a two-dimensional lattice; the height of the unit cell is the distance between lattice planes.

three-dimensional lattice; the height of the cell is thus the distance between the two-dimensional lattice planes (Figure 3.5).

In some of the cells shown in Figures 3.2 and 3.3 there are lattice points at face centers or cell centers, in addition to the vertices. Crystallographers use such 'centered' cells when this increases the symmetry of the cell. The volume of a centered cell is a multiple of the volume of a true unit cell. When centered cells are used, the lattice points cannot all be given integer coordinates with whole numbers. This wreaks havoc with the mathematical theory, so in this book 'unit cell' will always mean a true unit cell, unless explicitly stated otherwise.

3.3   The Voronoï Cell

In addition to the parallelepipedal unit cells, there is an alternative shape for the building blocks of a lattice. Like the unit cells, these cells contain one lattice point, but it lies at the cell's center. Unlike the unit cells, these cells are unique and their symmetry groups are the stabilizer groups of the lattice points. They are called the *Voronoï cells* of the lattice. (Voronoï cells are also known, *inter alia*, as Dirichlet domains, Brillouin zones, and Wigner–Seitz cells, reflecting the fact that they have been rediscovered in many different contexts; priority goes to Dirichlet.)

3.3.1   Constructing Voronoï cells

The Voronoï construction is quite general; it can be used for any discrete set of points, finite or infinite. To see how it works, we will construct the Voronoï cell of the point marked x in Figure 3.6(a). We first join this point to all the others by straight line segments (Figure 3.6(b)). Next, we construct the orthogonal bisector of each line segment (Figure 3.6(c)). The smallest convex region bounded by the bisecting lines is the Voronoï cell of the center point (Figure 3.6(d)). We can think of it as the region of the plane 'occupied' by x.

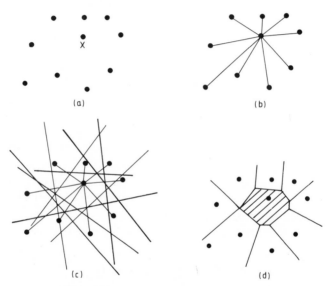

Figure 3.6. Constructing a Voronoï cell.

If we carry out the same construction for each point of the set in Figure 3.6(a), we obtain a partition of the plane into cells (Figure 3.7). Every point of the plane either lies inside such a cell, or on the edge shared by two of them, or at a vertex shared by three or more. There are no gaps between cells, the cells don't overlap, and they share whole edges. The points of the set whose cells share a vertex are equally distant from that vertex.

The same construction works for point sets in any dimension. For example, if x is a point in three-dimensional space, we join it

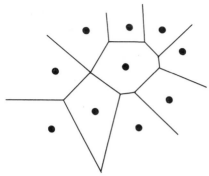

Figure 3.7. The Voronoï cells of the points in Figure 3.6(a).

to the other points of the set by straight line segments, and bisect these segments with planes orthogonal to them. The Voronoï cell of x is thus a polyhedron.

3.3.2 Voronoï cells of lattices

In any regular (r, R) system the configurations about each point are identical; thus the Voronoï cells of the points of a regular (r, R) system will also be identical. In particular, the Voronoï cells of the points of a two-dimensional lattice will be congruent polygons that fill the plane, and the cells of a three-dimensional lattice will be congruent space-filling polyhedra.

In any dimension, the Voronoï cell of a lattice point is always centrosymmetric. This means that if the lattice is two-dimensional, its Voronoï cells must have an even number of sides. Thus they must be either rectangles or hexagons (Figure 3.8) because, as we saw in Chapter 1, congruent polygons with more than six sides cannot tile the plane. In higher dimensions, the faces of the cell are also centrosymmetric, because each face contains a center of symmetry: the midpoint between a pair of lattices points.

There are five types of three-dimensional Voronoï cells, when the cells are classified combinatorially by the number and arrangement of faces, edges, and vertices; representatives of each type are shown in Figure 1.12. Fedorov, who discovered them, called them *parallelohedra* because they fill space when placed face to face in parallel position.

It is not difficult to figure out why there are only five types.

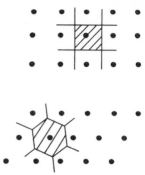

Figure 3.8. The Voronoï cells of plane lattices are rectangles or hexagons.

The simple argument we present here, originally due to Delone, is a challenging exercise in spatial visualization.

First we note that, for any three-dimensional lattice, the fact that the faces of the Voronoï cells are centrosymmetric means that they occur in bands defined by each pair of parallel edges: for any face and any edge of that face, there is an opposite edge shared with a second face; there is an opposite edge of the second face shared with a third face, and so forth. The band always closes on itself, because the cell has a finite number of faces (Figure 3.9).

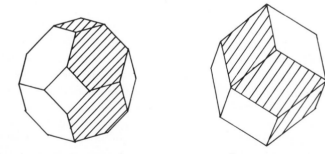

Figure 3.9. Because the faces of the Voronoï cell of a lattice point are centrosymmetric, they occur in closed bands.

The centers of the faces in each band lie in a plane which is orthogonal to the direction of the edge defining the band. This plane also contains the center of the cell. Since adjacent cells share whole faces, the centers of their faces coincide. This means that in the three-dimensional tiling corresponding to the decomposition of space into the Voronoï cells of a lattice, the cells occur in parallel layers (Figure 3.10).

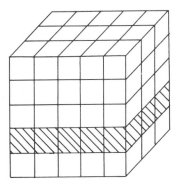

Figure 3.10. The Voronoï cells of a lattice occur in parallel layers.

A plane cross-section of such a layer, cut through the centers of the cells, is a Voronoï decomposition of a two-dimensional lattice, so it is a tiling of the plane by rectangles or by hexagons. The edges of these polygons are the lines in which the plane has cut the cell faces; the vertices of the polygons are the points in which the plane has cut the cell edges (Figure 3.11). Thus the edges of the cells in a two-dimensional cross-section correspond, in three dimensions, to a band of faces.

To find the shapes of the three-dimensional cells, we must reconstruct them from their cross-sections. We first consider the ways in which identical layers of two-dimensional cells can be superimposed on one another. The edges of a cell in one layer define a band of faces of the corresponding three-dimensional Voronoï cell; from the position of the edges, vertices, and faces of the cells of the other layers we can figure out the faces of the Voronoï cell above (or below) the band. The faces below (or above) the band are then also known, since the Voronoï cell is centrosymmetric.

We now carry out this program.

When the two-dimensional cells are rectangles, there are three distinct ways to superimpose the layers (Figure 3.12); when they

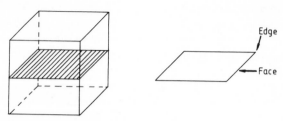

Figure 3.11. The edges of these polygons correspond to three dimensional cell faces, the vertices to cell edges.

are hexagons there are four ways (Figure 3.13). This suggests that there are seven parallelohedra, but, as we will see, there are some duplications on our list.

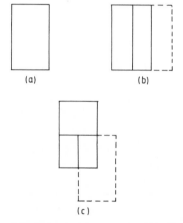

Figure 3.12. Three distinct ways to superimpose layers of rectangular cells.

The basic principles of deduction by superposition include the following.

1. If an edge of a cross-section polygon in one layer is fully superimposed on an edge in the layer below, then the faces containing these pairs of edges are rectangles.

2. If an edge of a cross-section polygon in one layer is superimposed on two edges, or crosses an edge, in the layer below, then

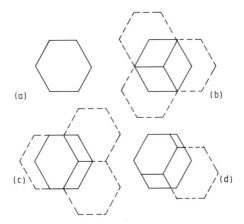

Figure 3.13. The four superpositions of hexagonal layers.

the corresponding faces are hexagons.

3. The decomposition of a cross-section polygon by the superposition of polygons in the layer below tells us the number and shapes of the faces that the cell shares with cells in the layer below.

Let us see how this works in the case when the cross-sections of our parallel layers are rectangles (compare Figures 3.12 and 3.14).

In Figure 3.12(a), the edges of the cross-sections are directly superimposed, so the faces of the band are quadrilaterals. Each of the cells in one layer shares a quadrilateral face with exactly one cell in the layer below. This is only possible if the cell is a rectangular solid (Figure 3.14(a)).

In Figure 3.12(b), two of the faces in the band are quadrilaterals and two are hexagons. Also, each cell shares faces with two cells in the layer below, and these faces are quadrilaterals. Thus the cell is a hexagonal prism (Figure 3.14(b)).

In Figure 3.12(c), all four faces of the band are hexagons, and the four faces which it shares with cells below are quadrilaterals. This cell (Figure 3.14(c)) has four hexagonal faces and eight quadrilateral ones (four above, four below). It is sometimes called an elongated dodecahedron.

Now you should be able to show that the cell in Figure 3.13(a) is a hexagonal prism, in (b) it is a rhombic dodecahedron, in (c) it is an elongated dodecahedron, and in (d) the cell is a truncated

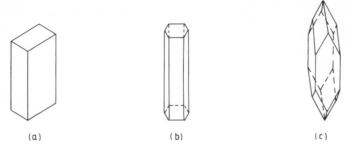

Figure 3.14. Reconstruction of the three-dimensional cells whose cross-sections are shown in Figure 3.12.

octahedron. Thus the seven possible kinds of cells are in fact five.

Figure 3.15 shows how the five Voronoï cells are apportioned among the 14 Bravais lattices. The relationship between Bravais lattices and Voronoï cells is not a simple one; in some cases, more than one combinatorial type of Voronoï cell can correspond to a single Bravais lattice when the metric specifications of the lattice depend on more than one parameter. For example, a tetragonal body-centered lattice (type I) with parameters a (in two directions) and c (in the other) has a truncated octahedron for its Voronoï cell if $c/a < \sqrt{2}$; if $c/a \geq \sqrt{2}$ the cell will be an elongated dodecahedron.

3.4   The dual lattice

We now return to the lattices themselves. Let Λ be a three-dimensional lattice generated by the vectors a, b and c. Since Λ is an orbit of the group generated by the corresponding translations, we can label each lattice point with the translation vector connecting it to the origin. Such a label will be of the form $ua + vb + wc$, or (u, v, w) if a, b and c are understood.

In the a, b, c coordinate system, a lattice plane through the origin has an equation of the form

$$hx + ky + lz = 0$$

where h, k, and l can be taken to be integers with no common factor.

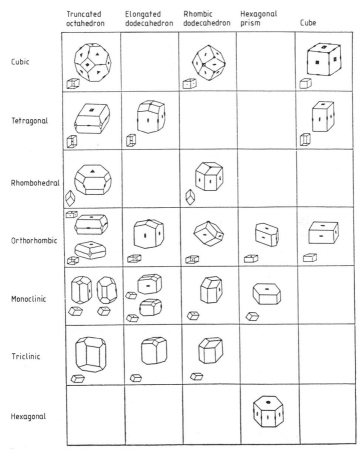

Figure 3.15. The distribution of Voronoï cell types among the Bravais lattices. Adapted from *A Bravais: Collected Scientific Works*, B N Delone and I I Shafronovskii, eds, Leningrad, Nauka, 1974 (in Russian).

Thus for any triple of integers (u, v, w),

$$hu + kv + lw = \text{an integer} \tag{3.1}$$

and (3.1) is the equation of a plane parallel to the first one. Every lattice point satisfies this equation, and every integer will appear on the right-hand side of (3.1) for appropriate choices of (u, v, w).

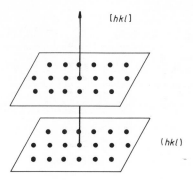

Figure 3.16. The lattice is decomposed into parallel planes.

Thus the lattice is completely decomposed into stacks of parallel planes by the choice of (h, k, l) (Figure 3.16).

In a cartesian coordinate system, the expression $hu + kv + lw$ is a dot product:

$$hu + kv + lw = (h, k, l) \cdot (u, v, w)$$

so we can interpret the ordered triple (h, k, l) as the *cartesian* coordinates of a vector normal to the stack of parallel planes. In this coordinate system (h, k, l) is, in general, not a lattice vector. Thus if we want to think of (h, k, l) as a linear combination of three basis vectors, the basis will not be a, b, and c but some other triple of vectors a^*, b^* and c^* which have the property that, for any integers u, v and w,

$$(ha^* + kb^* + lc^*) \cdot (ua + vb + wc) = hu + kv + lw.$$

This is only possible if

$$a \cdot a^* = b \cdot b^* = c \cdot c^* = 1$$

and

$$a \cdot b^* = a \cdot c^* = b \cdot a^* = b \cdot c^* = c \cdot a^* = c \cdot b^* = 0.$$

These relations completely define the vectors a^*, b^* and c^* (Figure 3.17). The integral linear combinations of these three vectors

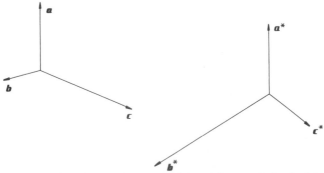

Figure 3.17. The lattice Λ generated by a, b, c, and the dual lattice Λ^* generated by a^*, b^*, c^*.

form a translation group whose orbit is again a lattice, Λ^*. In mathematics, Λ^* is called the *dual* lattice; in crystallography, it is the lattice *reciprocal* to Λ.

Equivalently, Λ^* can be defined as the set of all vectors (h, k, l) such that, for each (u, v, w) in Λ,

$$(h, k, l) \cdot (u, v, w) = \text{an integer}. \tag{3.2}$$

3.4.1 Geometry of the dual lattice

Equation (3.2) shows that each lattice is dual to the other: $\Lambda^{**} = \Lambda$. The vectors of Λ^* are orthogonal to the lattice planes of Λ, and the vectors of Λ are orthogonal to the lattice planes of Λ^*. In particular, the faces of the Voronoï cell of a point of Λ (chosen as origin) are parallel to lattice planes of Λ^*, and vice versa.

Although we are using the term 'dual' to be consistent with mathematical usage, the adjective 'reciprocal' is appropriate for several reasons. Consider, for example, the two successive planes of Λ whose equations are $hx + ky + lz = 0$ and $hx + ky + lz = 1$, and suppose d is the distance between these planes. Let (u, v, w) be a lattice point in the plane $hx + ky + lz = 1$. The length of the projection of the vector (u, v, w) onto the line containing the vector (h, k, l) is equal to d, and this projection is

$$\frac{(u, v, w) \cdot (h, k, l)}{|(h, k, l)|} = \frac{hu + kv + lw}{|(h, k, l)|} = \frac{1}{|(h, k, l)|} = d. \tag{3.3}$$

Thus the length of the vector (h, k, l) is reciprocal to the distance between the lattice planes orthogonal to the vector:

Theorem 3.1. $|(h, k, l)| = 1/d.$

Thus if we wish to find the stack of lattice planes with the largest interplanar spacing, we only need to find the shortest dual lattice vector. This is not difficult, since it will always correspond to a face of the Voronoï cell of Λ^*. Theorem 3.1 also tells us that, since there are arbitrarily large dual lattice vectors, there are arbitrarily small interplanar spacings in any lattice.

Since the unit cells of a lattice all have the same volume, and since the height of a unit cell is the distance between the lattice planes containing opposite bases of the cell, we conclude that

Corollary. $|(h, k, l)|$ *is proportional to the density of the plane lattice to which it is orthogonal.*

A lattice Λ and its dual lattice Λ^* always belong to the same crystal system, but they may belong to different Bravais classes. For example, in the cubic system the F and I lattice types are dual to one another, while the P lattice type is self-dual.

3.4.2 The dual lattice and diffraction

W H Bragg showed in 1915 that the diffraction of x-rays by crystals can be interpreted as reflection by the lattice planes of the crystal. When a beam of parallel, monochromatic x-rays of wavelength λ is passed through a crystal, the reflected rays will emerge from the crystal in phase if the wavelength λ, the interplanar spacing d, and the angle of reflection θ satisfy Bragg's condition:

$$n\lambda = 2d \sin \theta \qquad (3.4)$$

where n is an integer. If this condition is satisfied, and the emerging waves strike a photographic plate, they will create a pattern of bright spots. The x-ray crystallographer begins with these spots and works backward to deduce the geometry of the structure that gave rise to them.

The pattern of spots is closely related to the Fourier transform of the crystal's 'density function', which describes its distribution of mass. A general discussion of Fourier transforms and density functions is far, far beyond the scope of this book, but when we simplify the crystal structure to a point lattice, then its density function is very simple and so is its Fourier transform.

The density function $\rho(x)$ of a lattice Λ is a sum of Dirac delta functions located at the lattice points. (The Dirac delta function $\delta(x - a)$ is defined to be zero everywhere except at the single point a; its value at a is unbounded.) That is,

$$\rho(x) = \sum_{a \text{ in } \Lambda} \delta(x - a).$$

Neither the delta function nor any sum of deltas is a true function, but they can be given rigorous interpretations. They are useful tools because they allow us to use mathematical abstractions, such as point lattices, in physical theories.

The Fourier transform of $\delta(x - a)$ is $\exp(2\pi i t \cdot a)$. This is a complex quantity of modulus 1; it is equal to 1 when $t \cdot a$ is an integer. The Fourier transform of $\rho(x)$ is the sum

$$\hat{\rho}(t) = \sum_{a \text{ in } \Lambda} \exp(2\pi i t \cdot a).$$

Since each term of this sum is 1 when $t \cdot a$ is an integer, and since there are infinitely many points in Λ, $\hat{\rho}(t)$ will be unbounded whenever t is in Λ^*. We can also show that $\hat{\rho}(t) = 0$ for all other values of t. Thus $\hat{\rho}(t)$ is itself a sum of delta functions. It follows from this that the pattern of spots on the photographic plate is also a lattice, not Λ but Λ^*. This is why the dual lattice is fundamental in the interpretation of diffraction patterns.

Not all points of Λ^* will appear in a diffraction pattern. In part this is inherent in experimental technique, but there are also theoretical reasons. Using Theorem 3.1, equation (3.3) can be rewritten in the form

$$|(h, k, l)| = \frac{2 \sin \theta}{n\lambda}. \tag{3.5}$$

Thus the right-hand side of (3.4) must be equal to the length of a dual lattice vector. This implies that the dual lattice vectors that appear are limited to a sphere of radius $2/\lambda$; in crystallography this is known as the Ewald sphere.

3.4.3   *The dual lattice and crystal form*

The dual lattice is also a key to the relation between crystal structure and crystal form. As Figure 1.4 suggests, Haüy developed his building-block theory in order to explain this relation. After his blocks were abolished and replaced by systems of points, the question arose again. Bravais devoted a great deal of thought to this matter. He summarized his conclusion in the 'Law of Reticular Density', which states that:

The faces that appear on a crystal will be parallel to the lattice planes with the greatest density.

The 'law' holds remarkably well although, since real crystals are not point lattices, there are important exceptions (we will discuss one type of exception in Chapter 6).

The reasoning behind Bravais' law can be summarized in the following way. As we have seen, the density of a plane lattice is the area of a plane unit cell. Since the volume of a cell is independent of the choice of edges, and since that volume is equal to the area of the plane cell times the distance between successive lattice planes, the smaller the area of the planar cell the larger the interplanar spacing, and vice versa.

We now assume that in a simple crystal the surface free energy of a lattice plane is inversely proportional to the density and thus proportional to the interplanar spacing. If we also assume that the surface free energy is proportional to the growth rate of the crystal in the direction orthogonal to the lattice plane, then large interplanar spacings correspond to rapid growth, and small interplanar spacings to slow growth.

Paradoxically, the faces that appear on a crystal are parallel to the lattice planes with the slowest growth rates: the faster growing directions tend to 'grow themselves out of existence' (Figure 3.18). Hence Bravais' law.

It was G Wulff, already famous in crystal growth theory for his derivation of the equilibrium form, who observed in the early 1900's that Bravais' law actually predicts more: it tell us the ideal crystal form.

Wulff's equilibrium form is given by the following construction. From a point inside a crystal, draw vectors in all directions, with lengths proportional to the free surface energies of the possible face planes orthogonal to them. At the tip of each vector, construct an orthogonal plane. The equilibrium form of the crystal is the

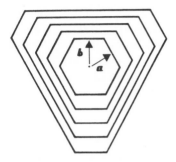

Figure 3.18. The growth rate in direction a is greater than in direction b; the a faces will disappear.

smallest convex region bounded by these planes.

Although the Wulff construction presupposes nothing whatsoever about the internal structure of the crystal, it can be applied to a crystal with a lattice structure, which Wulff did. Choosing the interior point to be a lattice point and assuming that the surface free energy of a lattice plane is inversely proportional to its density, he constructed vectors orthogonal to the lattice planes whose lengths were directly proportional to their densities, and constructed orthogonal planes at their tips. In other words, he constructed dual lattice vectors and planes orthogonal to them. The Wulff form for a lattice Λ is thus identical to the Voronoï cell of Λ^*, except for scale.

This means, for example, that a crystal with a primitive cubic lattice should have the form of a cube, a crystal with a face centered cubic lattice should have the form of a truncated octahedron, and a crystal with a body-centered cubic lattice should have the form of a rhombic dodecahedron.

In the tetragonal case, in which the primitive lattice axes are orthogonal with ratios $c : a : a$, a crystal with a primitive lattice should look like a rectangular solid; a crystal of the I type should have the form of a truncated octahedron (if $c/a > 1$) or an elongated dodecahedron (if $c/a < 1$). (It is *not* always the case that if Λ has a Voronoï cell of one type, then Λ^* has the other; when $1 < c/a < \sqrt{2}$, both cells are truncated octahedra.)

As is the case with Bravais' law itself, agreement with reality is surprisingly good, especially when you consider that this construction ignores the chemical composition, the actual arrangement of (real) atoms, and the environmental conditions of growth.

As an interesting postscript to this discussion, it turns out that the Wulff form can also be predicted by that most modern of mathematical models, the cellular automaton. In a series of papers in the late 1970's S Willson studied finite configurations of points located at the nodes of a lattice and subject to rules which state how additional points are to be added to the configuration. These rules are 'local' in the sense that whether and where points are added in the neighborhood of a given point depends only on the location of other nearby points. Willson showed that *any* initial configuration above a certain threshold size will converge to a polyhedral form which is (1) independent of the initial configuration, (2) is classically crystallographic, and (3) is essentially the Wulff form. This implies that once started, a crystallite will grow (at least theoretically) into a perfect periodic crystal, repairing any accidental imperfections of shape and achieving the form predicted by Bravais.

Chapter 4

The Space Groups

4.1 Chronology

The 230 space groups, so numerous and mysterious, are the symmetry groups of regular systems of points. They combine the point group of the crystal with its lattice structure to form the symmetry group of the system as a whole. The space groups have a long and interesting history, which is reflected in the fact that, like the Voronoï cells, they are known by many different names: space groups, crystallographic groups, Fedorov groups, Bieberbach groups, and probably others as well.

We have seen that lattices were introduced by Haüy's critics in about 1824, and that the lattices were classified correctly according to their symmetry by Bravais in 1850. Bravais' work inspired C Jordan to study regular systems of points more general than lattices: orbits of groups of 'motions'. By motions Jordan meant the orientation-preserving isometries, sometimes called proper motions: rotations, translations, and screw rotations. His study of these groups was published in 1868.

Jordan did not enumerate any groups; this was carried out by L Sohncke in 1879. Later it was realized that reflections, glide reflections, and rotary reflections should be included in the groups. Finally, the space groups were enumerated in 1891 by Fedorov and Schoenflies, independently though cooperatively. This was nearly 70 years after the pioneering work of Haüy, and about 21 years before the discovery of x-ray diffraction by crystals in 1912. Coincidentally (in both senses of that word), the basis for the theory of N-dimensional space groups was laid by G Frobenius and Bieberbach at about the same time as that discovery.

Throughout the long history of its development, until diffraction photographs were available to give powerful experimental support for the lattice hypothesis, only a very few people believed that the space groups were anything more than an arcane, rather complicated geometrical exercise. But by 1915, the importance of the space groups for x-ray studies had been recognized, and the groups had been dusted off and recast in a form that could be used by crystallographers. In 1978, after the advent of the high-speed computer, the 4901 four-dimensional space groups were enumerated by a team of mathematicians, computer scientists, and crystallographers.

This unfolding sequence of generalizations reflects an ongoing debate about the geometry of crystalline order. The debate is an old one: W Barlow and H Miers wrote in 1901,

> *The history of its development ... is the history of an attempt to express geometrically the physical properties of crystals, and at each stage of the process an appeal to their known morphological properties has driven the geometrician to widen the scope of his inquiry and to enlarge his definition of homogeneity.*

This is precisely the reason why further generalizations are being explored today.

There is a puzzling gap in this chronology. Even before Jordan's study of groups of 'motions', it should have been evident that reflections and glide reflections needed to be included among the symmetries of regular systems of points. L Pasteur's discovery, in the early 1850's, of the relation between molecular and morphological chirality should have made this clear both to experimental crystallographers and to their more theoretical colleagues. Yet the necessary generalization was incorporated into space group theory only much later. This is an interesting question for historians of science to explain. Was it due to differences in scientific fields, or differences in language? Are there similar gaps today of which *we* are unaware?

The 87 year gap between the enumeration of the three- and four-dimensional space groups was even greater, but this one is easier to explain. The time lag was due not only to the ferocity of such a computation, but also to the lack of a conceptually simple algorithm and to the absence of any need for these groups. The necessary algorithm, based on the equation of Frobenius that is the central theme of this chapter, was supplied independently by J J Burckhardt and by H Zassenhaus. The need for four-dimensional space groups was supplied in the early 1970's when, as we have already noted, crystallographers began to use them. With computers available to implement the algorithm, the necessary tools were in place for the task of enumeration.

The algorithm works for space groups in any dimension and so does the computer so, in principle, the space groups in any dimension can now be enumerated. But the number of groups grows astronomically with increasing dimension, and no one is likely to have the computer time or professional interest to extend this counting any further. It is more likely that special classes of higher dimensional groups will be studied in relation to particular problems (see Chapter 7).

4.2 The orbits of a space group

Since they are the symmetry groups of regular systems of points, space groups are infinite groups of isometries whose orbits are discrete and relatively dense. Theorem 1.1 tells us that any space group G has a subgroup T generated by translations. It is the largest subgroup of G which is commutative; this property can be used to define the space groups.

Another way to characterize the space groups is to say that they are translation groups 'extended' by point groups. Every orbit of a space group is a union of a finite number of congruent lattices (Theorem 1.2); each of the lattices is an orbit for the translation group. The symmetry operations associated with the point group permutes the lattices; different point operations affect different permutations. A very simple way to visualize these permutations is to begin with a generic orbit and assign the points of each lattice a different color. The translations map each colored set onto itself; the other symmetry operations permute the colors. For example, Figure 4.1 shows a generic orbit of the symmetry group of the square lattice which has been 'colored' in this way. (The standard notation for this group is p4m.) We see that the orbit is a union of eight different lattices, related to one another by rotations and reflections. Every symmetry operation in the group effects a color permutation. For example, the eight colors, in two sets of four, are permuted cyclicly by 90°, 180°, 270° and 360° rotations about the point marked x.

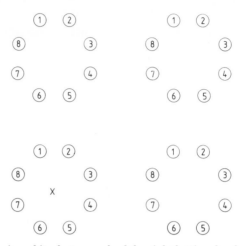

Figure 4.1. An orbit of p4m; each of the eight lattices has been assigned a number representing a color.

The symmetry operations never mix the colors: if one 'red' point is mapped to a 'blue' one, then every 'red' point is mapped to a 'blue' one. This is because an isometry must map a lattice to a congruent copy of itself, i.e., another lattice. Since all points of each lattice have the same color, the isometry maps colors to colors unambiguously.

Although orbit diagrams are helpful in understanding how a space group works, they are not a very efficient means of enumerating all the space groups. For that, we need to do some algebraic computations. And to carry out the algebraic computations, we need to know more about subgroups.

4.3   Cosets and normal subgroups

Let G be a group and H any subgroup of G. H partitions the group G into sets called *cosets* in the following way. If H is the set of elements

$$e, h_1, h_2, \ldots$$

and if g is an element of G which is not in H, then the multiples

$$g, gh_1, gh_2, \ldots$$

are all different elements of G. We denote this set of multiples by gH; it is called a *coset* of H, with coset representative g. (Any element of a coset can serve as the coset representative.) We regard H itself as a coset with representative e.

If there are any elements left in G which are not in H or in gH, we can choose one of them and form a new coset with it. Again the elements of this coset are different from one another; they are also different from the elements of the first coset. We can continue forming cosets until every element of G belongs to one. In this way, H partitions G into sets of the same cardinality as H. The number of sets (counting H too) is the index of H in G. Lagrange's Theorem (Chapter 2), which states that if G is a finite group, then $|G|/|H|$ is a whole number, is an immediate consequence of the partitioning of G by H and its cosets.

In constructing these cosets, we multiplied the elements of H by g on the left: the set gH is actually a *left* coset. We can also partition G into *right* cosets; everything we have said still applies. In general, the left cosets and the right cosets of a subgroup are not the same unless G is a commutative group.

For example, let G be the symmetry group of the square. This group, denoted 4m, has eight elements: the identity, rotations r, r^2, r^3 through $90°, 180°, 270°$, reflections d_1 and d_2 in the diagonals of the square, and reflections b_1 and b_2 in the two lines which bisect its edges (Figure 4.2).

The inclusion relations among the subgroups of 4m are shown in Figure 4.3. There are eight proper subgroups, three of order four and five of order two.

A subgroup of order four has only one coset apart from itself, so it is both a left and a right coset (Figure 4.4).

The left and right cosets of the subgroups of order two, on the other hand, are not the same in four of the five cases (the exception is the subgroup generated by r^2). The subgroup generated by d_1 is typical of the others. Its left cosets are shown in Figure 4.5; its right cosets are shown in Figure 4.6.

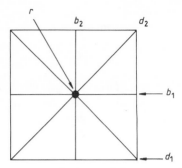

Figure 4.2. The symmetry elements of the group of the square, 4m.

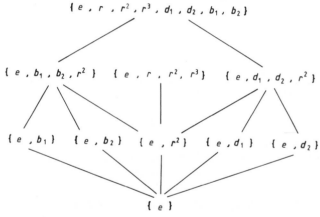

Figure 4.3. The subgroups of 4m.

When the right and left cosets of a subgroup are identical, then for each g in G we have $gH = Hg$. This means that $gHg^{-1} = H$: H is a *normal* subgroup.

The left, or what is the same thing, the right cosets of a *normal* subgroup form a group themselves: the normal subgroup plays the role of the identity element, and the multiplication rule is:

$$(gH)(g'H) = g(Hg')H = g(g'H)H = gg'H. \qquad (4.1)$$

Equation (4.1) shows that the set of cosets is closed.

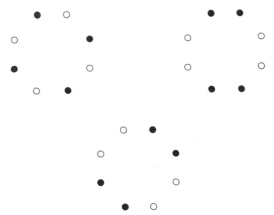

Figure 4.4. Orbits for the three subgroups of 4m of order four. The filled circles are orbits for the subgroups; the empty circles represent the cosets (or vice versa).

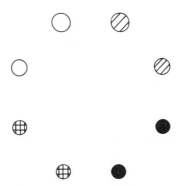

Figure 4.5. The dark circles are an orbit for the subgroup of 4m generated by d_1. The light, empty, and hatched circles represent the *left* cosets.

The group of cosets of a normal subgroup H is called the *quotient group* of G by H, and is denoted G/H. The order of G/H is the index of H in G.

As a first example, let us compute the quotient groups G/H for $G = 4m$. 4m has four normal subgroups, the three subgroups of order four and the subgroup of order two generated by r^2. When

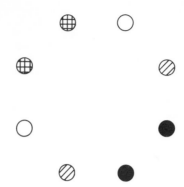

Figure 4.6. The dark circles are the same as in Figure 4.5, but now the light, empty, and hatched circles represent the *right* cosets.

$|H| = 4$, G/H is a group of order two; when $|H| = 2$, G/H is a group of order four. There is only one group of order two, but there are two different groups of order four: one is cyclic, the other is not. Which kind is G/H? It will be cyclic if the group of cosets is generated by a single element. To find out whether it is, we use equation (4.1). The cosets of H are rH, d_1H, and b_1H. We have: $(rH)(rH) = H$, $(d_1H)(d_1H) = H$, and $(b_1H)(b_1H) = H$. Thus the four cosets are not powers of a single one of them; G/H is not cyclic.

Quotient groups of infinite groups work the same way: here is a simple example. Let T be a one-dimensional lattice, generated by a single translation t. Let G be the group generated by t and by p, a rotation of 60° about an axis parallel to the direction of translation (Figure 4.7).

Then $p^6 = e$ and

$$G = T \cup pT \cup p^2T \cup \ldots \cup p^5T.$$

Since rotations and translations along the rotation axis commute, T is a normal subgroup of G. Using equation (4.1), we find that

$$(pT)^6 = p^6T = eT = T,$$

so G/T is isomorphic to the cyclic group of order six generated by p.

Figure 4.7. Part of an orbit for the group generated by a translation and a rotation.

The next example is slightly more complicated, but it is helpful for understanding the space groups. Let G be the group generated by the same translation group T and, this time, a screw rotation p: a rotation through 60° together with translation through $t/6$. Now $p^6 = t$ (Figure 4.8). Again p and t commute, so T is normal in G. Notice that, since t is a translation, we again have $(pT)^6 = T$, so the two quotient groups we have constructed are isomorphic.

Figure 4.8. Part of an orbit for the group generated by a translation and a screw rotation.

From our coloring argument in the preceding section, we see that the translation subgroup T of a space group G is a normal subgroup. The quotient group G/T is always finite, since the number of lattices in a space group orbit is finite. This finite group is isomorphic to a crystallographic point group P. When we pass from G to G/T and thus to P, we are, in effect, ignoring the translations of G.

The examples above show that the same point group can correspond to space groups with different properties.

4.4   Constructing the space groups

To construct a space group G, we combine a translation group T and a point group P. The group G that we build in this way will have the property that the quotient group G/T is isomorphic to P. But in order to distinguish among different groups with the same T and P, we need to decide how multiplication in G will work. That is, we need to know how the elements of P transform the lattice vectors. There are usually several different ways that P can do this, and consequently there can be several different space groups with the same lattice and point group.

We will discuss just one class of two-dimensional space groups, all with the same point group 4m; this will be sufficient to demonstrate the basic principles of space group construction. Our purpose is not to enumerate the groups but to understand the ideas that make enumeration possible. It turns out that all the space groups, in any dimension, can be derived from one simple equation!

First we introduce some notation especially designed for space groups. It is customary to use additive notation (+) for translations and multiplicative notation (juxtaposition) for point operations. It would be nice to have a symbol for multiplication in G that allows us to use both notations at the same time. To this end, elements of the space group G will be denoted by ordered pairs whose first components are added and whose second components are juxtaposed.

The identity element will be $[0, e]$. We 'embed' T in G by associating t with $[t, e]$. Then $[t_1, e] + [t_2, e] = [t_1 + t_2, e]$. The set of all elements $[t, e]$ is then a subgroup of G, which looks and acts just like T.

But we need to handle P a little differently so that a rotation or reflection in P may take the form of a screw rotation or glide reflection as an element of G. To do this, we associate each p in P with an element of G of the form $[s, p]$, where s may be 0 or may be a translation *not* in T. (When $p = e$, $s = 0$.) These elements of G multiply according to the rule

$$[s_1, p_1][s_2, p_2] = [s_1 + p_1 s_2, p_1 p_2].$$

If $p_1 p_2 = p_3$ in P, then

$$[s_1, p_1][s_2, p_2] - [s_3, p_3] = [t, e]$$

where t is a lattice translation (possibly but not necessarily 0). This implies the fundamental relation

$$s_1 + p_1 s_2 - s_3 = t \qquad (4.2)$$

which is sometimes written in the form

$$s_3 \equiv s_1 + p_1 s_2 \pmod{T}$$

and called the *Frobenius congruence*.

Equation (4.2) must hold for all space groups, whatever their dimension. If we solve it for every pair of elements of P, its solutions give us all space groups with point group P. But that is inefficient; we only need to solve it for the generators of P.

Turning to our two-dimensional example, in which $P = 4m$, let us choose as generators r and b_1.

We first consider the case when $p_1 = p_2 = r$; (4.2) becomes $s_1 + r s_1 - s_3 = t$. But it turns out that there is nothing to calculate: we know that $s_1 = s_3 = 0$, because there are no screw rotations in two-dimensional groups.

But two-dimensional groups do contain glide reflections, so if $p_1 = p_2 = b_1$ is a reflection, we have some work to do. First notice that since $b_1^2 = e$ in P, we must have $s_3 = 0$. So the equation we must solve is

$$s_1 + b_1 s_1 = t \qquad (4.3)$$

where t is a lattice translation.

Assume that the line across which b_1 reflects is the x-axis. As a vector in the plane, s_1 has coordinates (u, v), say. Then $b_1 s_1 = b_1(u, v) = (u, -v)$, and the equation becomes

$$(u, v) + (u, -v) = (2u, 0) = t.$$

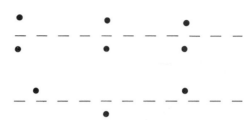

Figure 4.9. Reflection and glide reflection correspond to the two solutions of equation (4.3).

Either $u = 0$, in which case $t = 0$ and $[0, b_1]$ is a reflection, or else $(u, 0)$ is half a lattice translation and $[s_1, b_1]$ is a glide reflection (Figure 4.9).

Thus we see that when the point group P is 4m, we can construct two different kinds of two-dimensional space groups: with or without glide reflections.

To be complete, we should solve the equation for $p_1 = r$ and $p_2 = b_1$ (and vice versa), but the pattern is already clear. Further calculations show—we omit details—that there are exactly two groups (up to isomorphism—but see Chapter 6 for more about space group classification). Generic orbits for both groups are shown in Figure 4.10. The orbit in (a) is the same as the orbit in Figure 4.1.

If P is a cyclic group of order six generated by p and T is a one-dimensional translation group generated by the translation t, then it follows from equation (4.2) that, in G, p takes one of the forms $[0, p]$ or $[t/6, p]$. That is, p is a rotation or a screw rotation, as in the examples we discussed in the previous section.

4.5   Symmorphic and nonsymmorphic space groups

When all the generators of a space group can be chosen so that $s = 0$, the group is said to be *symmorphic* (in crystallography) or a *semi-direct product* (of its translation and point groups) or a *split extension* (in mathematics). It is easy to recognize an orbit of a

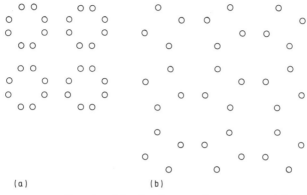

(a)   (b)

Figure 4.10. Generic orbits for the two plane groups with point group 4m. (a) p4m; (b) p4g.

symmorphic group: it consists of infinitely many copies of orbits of the point group. The orbits are copied over and over again by translations. Figure 4.1 (also 4.10(a)) is an example of such an orbit. So is every lattice: in this case the point group orbit consists of a single point.

The orbit of p4g shown in Figure 4.10(b) does not consist of copies of orbits of 4m. p4g is a typical *nonsymmorphic* group.

4.6 Subgroups of space groups

The subgroups of space groups are very important in crystallography. They describe hierarchical relations among crystal structures, they define allowed structural changes in some kinds of phase transitions, and, as we will see in Chapter 5, they play a key role in color symmetry. Another way to explain the difference between the orbits of symmorphic and nonsymmorphic groups is to say that symmorphic groups have subgroups isomorphic to their point groups, while nonsymmorphic groups do not.

The subgroups of space groups which are of greatest interest are those of finite index. In three dimensions, such a subgroup H must contain three linearly independent translations, so it is a space group too. Its translation subgroup T_H is a subgroup of

G's translation group T, and its point group P_H is a subgroup of G's point group P. The construction of subgroups is in every way analogous to the construction of space groups.

Historically, two kinds of subgroups have received special attentions: those for which $P_H = P$ and those for which $T_H = T$. (If both equalities hold, then $H = G$.) When $P_H = P$, H and G are said to be *class-equivalent*; when $T_H = T$, they are said to be *translation-equivalent*. A useful theorem of C Hermann says

Theorem 4.1. *Every subgroup of a space group which is of finite index is a translation-equivalent subgroup of a class-equivalent subgroup.*

Of course, this theorem is only true if we include improper as well as proper subgroups.

For example, the filled circles in Figure 4.11 represent an orbit of a subgroup H of G =p4m of index two. Here $P_H = P$ but T_H is a proper subgroup of T. Thus H is a class-equivalent subgroup of G (which is a translation-equivalent subgroup of itself).

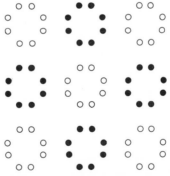

Figure 4.11. This subgroup of p4m of index two is class-equivalent.

In Figure 4.12 we see an orbit of a subgroup of p4m of index four. Here P_H is a proper subgroup of P and T_H is a proper subgroup of T. H is a translation-equivalent subgroup of the subgroup shown in Figure 4.11.

Symmorphic groups can have nonsymmorphic subgroups, and vice versa.

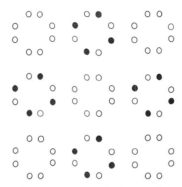

Figure 4.12. This subgroup is a translation-equivalent subgroup of the subgroup shown in Figure 4.11.

A subgroup is *maximal* if it is not contained in any larger (proper) subgroup. Theorem 4.1 says that every maximal subgroup of a space group G is either class-equivalent or translation-equivalent to G. These subgroups are relatively easy to find, and once we have found them, we can look for *their* class-equivalent and translation-equivalent subgroups. This is one way to generate lists of subgroups of space groups. Another method is to do an 'index search'. The index of a subgroup H in G is the product of the indices of P_H in P and T_H in T. So, for example, we can find all subgroups of index three by testing all index combinations 3×1 and 1×3. The first method is well-suited for finding chains of subgroups; the second is especially useful in color symmetry.

Chapter 5

Color Symmetry

5.1 Why colors?

In Chapter 1 we discussed several 'concrete' models of crystal structure, together with the more abstract model of (r, R) systems. Each of these models gives us insight into some of the properties of crystal structures; that is why we study them. But the attractiveness of sphere-packings, tilings, and networks (and point sets too) goes beyond their utility. Properly realized, they are *beautiful*. Repetition of parts by symmetry creates patterns that delight us—who does not enjoy looking through a well-made kaleidoscope, or at a medieval stained glass window? The pleasure that we get from repeating patterns is evidently very deep: they are part of the ornamental art of all cultures. The beauty of science is often thought to be of an abstract nature, but in crystallography it is both abstract and visual, and so doubly compelling. I strongly suspect that this is one of the reasons why the repeating-pattern paradigm of crystal structure has become so deeply rooted.

Many ornamental patterns are beautifully colored—the motifs are assigned different colors in ways that we perceive as symmetrical and harmonious. Colors serve to distinguish background from foreground in otherwise confusing patterns. They enable us to grasp the features of their subpatterns quickly, and to remember the relation of parts to the whole. This is perhaps the most important use of colors in scientific models.

Colored patterns are widely used in crystallography. For example, different colored spheres are often used to represent different kinds of atoms in ball-and-stick models of molecules and crystal structures. In the 1920's, Hermann used colors to describe repeating patterns in a two-sided plane: black points represented the points on one side of the plane, white points those on the other (Figure 5.1).

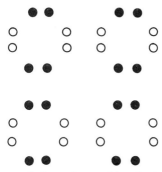

Figure 5.1. A two-coloring of an orbit of p4m. Black points are supposed to lie on one side of the plane, white points on the other.

Later, in the 1950's, these two-colored patterns were reintroduced by A V Shubnikov to describe arrangements of magnetic spins in crystal structures. Assuming that the spins of each kind form identical patterns, related to one another by symmetry operations, all the possibilities for spin distribution could be determined by systematically investigating symmetrically colored patterns. Figure 5.1 can be interpreted as one possible way of distributing spins in a p4m structure; Figures 4.11 and 5.2 show others.

Shubnikov's colleagues experimented with ways of coloring patterns with more than two colors. But mathematicians and crystallographers began to look for a coherent theory of color symmetry with an arbitrary number of colors only after they were introduced to M C Escher's graphic art. His intricate and beautiful colored interlocking designs raised puzzling and fascinating mathematical questions (Figure 5.3).

Escher endowed his various wiggly creatures with two, three, or four colors; their distribution was always symmetrical and harmonious. Clearly he had a method, but what was it? This is

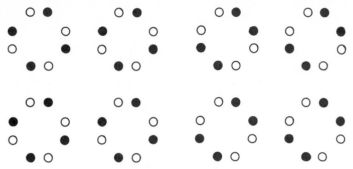

Figure 5.2. Two more symmetrical distributions of two colors in an orbit of p4m.

Figure 5.3. A colored tiling of the plane by Escher. ©1990 M C Escher Heirs / Cordon Art—Baarn—Holland.

a much more interesting question than was realized only a few years ago. Mathematically inclined scientists, including crystallographers, looked for classifications in which the coloring was governed, implicitly or explicitly, by symmetry groups. Escher did not use group theory, either implicitly or explicitly. Like artists in other times and cultures, he was interested in the way that neighboring tiles were related to one another; unlike most artists, he worked out a local theory of coloring requirements. We have seen (Theorem 1.3) that a requirement of local order can produce a regular pattern, but Escher's requirements were formulated rather differently. A discussion of his methods is beyond the scope of this book. However, the group-theoretic *interpretation* of Escher's patterns is not. It is the group theoretic interpretation of color symmetry that I will now describe. We begin with finite patterns.

5.2   Coloring finite figures

In this section, we will restrict ourselves to finite figures, but we will use an arbitrary number of colors (actually we will see that the number must be a divisor of the order of the symmetry group of the figure). For simplicity—and this is a very helpful simplification— we will assume that the figure is divided into its smallest equivalent parts, such as the sectors shown in Figures 2.2 and 2.3. Only one color can be assigned to each sector; adjacent sectors may or may not be given the same color.

If we were to distribute two colors among the sectors any which way, we would have a lot of possibilities: for the plate there are

$$\binom{12}{0} + \binom{12}{1} + \cdots + \binom{12}{12} = 2^{12}$$

ways to assign them, and there are 2^{48} ways to assign two colors to the sectors of a cube (Figure 5.4 shows some of them). Of course, some of these many colorings differ only in orientation.

But if we include all possible colorings, we have no theory; we need some reasonable restrictions on the colorings we are willing to consider. It is generally agreed that the most reasonable restriction is the one used by Shubnikov and others: the sectors of each color should form identical patterns, each related to the other by symmetry operations. Such colorings are called *perfect*. This is indeed restrictive: there are only a few perfect two-colorings of the plate and the cube (Figure 5.5).

Figure 5.6 shows the only perfect three-coloring of the cube. (Are there any perfect three-colorings of the plate?)

The reason we are assigning colors to the smallest equivalent sectors of these figures is that each of these regions contains exactly

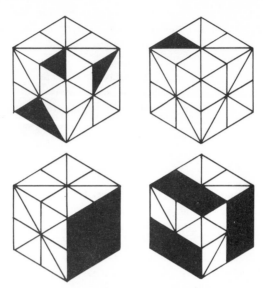

Figure 5.4. Four of the 2^{48} ways of coloring the sectors of a cube.

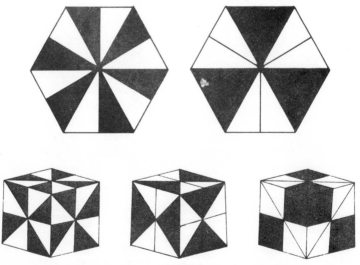

Figure 5.5. The perfect two-colorings of the plate and the cube.

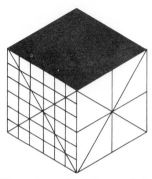

Figure 5.6. The only perfect three-coloring of the cube.

one point of any generic orbit of the symmetry group of the figure. We can say that the sectors themselves constitute a generic orbit.

Generic orbits are especially easy to work with in coloring theory, because they enable us to 'see' all the subgroups of the symmetry group. Because their stabilizers are trivial, every image gx of a point x must be different from x. It follows that if $g \neq g'$, then $gx \neq g'x$. This means that there is a one-to-one correspondence between the points of a generic orbit and the elements of the symmetry group. So, if we choose any point to be our starting point x, we can label x with the identity e and each of the other points of the orbit by the name of the symmetry operation that maps x to that point. Figure 5.7 shows how this can be done for a generic orbit of 4m; notice that this labelling is implicit in Figures 4.4 and 4.5.

Now we can use group theory to determine all the perfect colorings of any figure. Let us take the cube as an example. We assume that it has been subdivided into its 48 sectors, but no colors have been assigned to them. Someone hands us jars of black and red paint; this is to be a two-coloring. How shall we choose the black sectors? The first thing we note is that our requirement that the sets of colored sectors form congruent patterns means that the number of sectors we color must be a divisor of 48. More: the requirement that the colored sets be interchangeable by symmetry operations of the cube group G implies that the sets form an orbit of some sort: the elements of G must permute the colored sets. There are two possible permutations of black and red: the colors

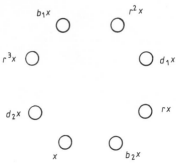

Figure 5.7. Labelling the points of a generic orbit of 4m (see Figure 4.2).

may be interchanged, or they may be kept fixed. Now we invoke the closure of G: since the resultant of two successive interchanges is the identity permutation, there must be elements of G which preserve the colors. Since the resultant of two color-preserving operations is another one, the set of elements of G which preserve the colors is a subgroup H. The remaining elements of G, all of which interchange the colors, constitute the coset gH.

Thus, to paint the cube perfectly with two colors (Figure 5.5), we select one of the three subgroups H of index two. Then, since we know how to label the sectors with group elements, we find all the sectors labelled with elements of H and color them black. Then we color the remaining sectors red.

Here is the general procedure, for k colors. We find a subgroup H of index k, and find the sectors which are labelled with elements of this subgroup. This is an orbit for H. We then paint these sectors with color 1. Next, we choose a symmetry operation g which is not in H, look for the sectors labelled with elements of the *left* coset gH, and paint them with color 2. (These coset sectors form a configuration congruent to the configuration of sectors with color 1; all we have to do is to see how g transforms those sectors. The image of the color 1 set under g is the color 2 set.) If H is a subgroup of index two, then we are done. Otherwise, there will be some sectors which have not yet been colored. We choose another element of G, not in H or gH, and identify the sectors associated with the corresponding left coset. We continue this procedure until all the sectors are colored. The number of colors is of course the index of H in G. Figure 5.8 shows each stage of coloring a square with four colors.

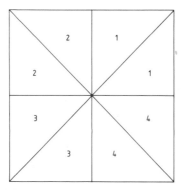

Figure 5.8. Coloring the sectors of a square with four colors.

This procedure for coloring a figure shows that each symmetry operation in G is associated with a unique color permutation. For, we have constructed a one-to-one correspondence between the sets of colored sectors and the cosets of H:

$$\text{color } 1 \leftrightarrow H, \quad \ldots \quad \text{color } k \leftrightarrow g_k H.$$

Since, for each g' in G, the set $g'(gH)$ is another coset of H, when the (left) cosets of H are multiplied (on the left) by g' they are permuted. This is, in effect, a permutation of the colors.

The group of the cube has subgroups of orders 24, 16, 12, 8, 6, 4, 3, 2, and 1, and so there are perfect colorings of its sectors with 2, 3, 4, 6, 8, 12, 16, 24 and 48 colors. In some cases there is more than one subgroup of a given order; when, as in the case of two colors, the subgroups are not conjugate, we get very different colorings. This is the situation in Figure 5.5.

But it is important to note that although the subgroup H specifies how the symmetry operations of G are associated with color permutations, it does not completely determine what the coloring will look like. This depends on the orbit that we choose for H. Figure 5.9 shows another coloring of the square with four colors which represent *the same* assignment of color permutations to symmetry operations as the coloring in Figure 5.8.

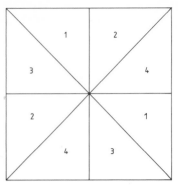

Figure 5.9. This coloring is defined by the same subgroup H as the coloring in Figure 5.8.

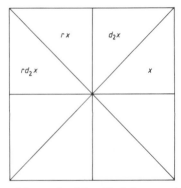

Figure 5.10. $\{r, rd_2\} = r\{e, d_2\}$; this left coset is associated with a copy of an orbit of H.

The reason for this strange state of affairs lies in the difference between left and right cosets. In abstract group theory, one can work either with left or with right cosets, but as far as generic orbits are concerned, they are not interchangeable. If x is any point, and H any subgroup, then Hx is the set of images of the point x as it is transformed by the elements of H; that is, it is an orbit of H. A *left* coset gH corresponds to a copy of this orbit; it need not be an orbit itself (Figure 5.10). On the other hand, since for each g, gx is a point different from x, the set Hgx corresponding to the *right* coset Hg is another orbit of H. It need not be congruent to the first orbit (Figure 5.11).

H has as many incongruent orbits as it has conjugates. Thus, when H is a normal subgroup, it has only one kind of orbit, but otherwise it has several, and the differences in their colorings can be quite dramatic.

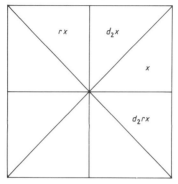

Figure 5.11. The right coset $\{r, d_2r\}$ is associated with another orbit of H.

5.3 Coloring infinite patterns

Coloring theory for infinite patterns is exactly the same as for finite ones, but it is more fun and it is more instructive.

We will start our discussion with Figure 5.3. Because Escher preferred to draw animals and people who appeared to be moving, his shapes are usually asymmetrical, as they are in this case. Thus, if we ignore colors for the moment, the lizards are a generic orbit for the symmetry group G of the uncolored pattern.

In order to include a larger portion of this orbit, we represent it schematically in Figure 5.12. Each marked hexagon represents a lizard; the numbers represent their colors. The pattern is perfectly colored.

Notice that the marks on the hexagons occur in three orientations; this tells us that the point group of G is a group of order three (there is only one group of order three, and it is cyclic). Notice also

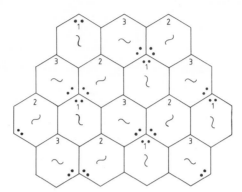

Figure 5.12. A schematic representation of Figure 5.3.

that this group is symmorphic: we can see this because the three orientations are grouped together and repeated by translations.

All of the marked hexagons of the same orientation have the same color. Let the hexagons with color 1 be the orbit for our subgroup H. We see that H contains only translations. In fact, it contains all of the translations of the group G, because every marked hexagon in this orientation has color 1. Thus H is a translation-equivalent subgroup of G, and $P_H = \{e\}$.

Since there are three colors, H has two cosets. In this example, H is a normal subgroup: we can see this because all three colored sets are preserved by translation and thus are orbits for H. This is an example of the type of coloring we used in Chapter 4 to distinguish the constituent lattices of an orbit of a space group. Notice also that every rotation of G is associated with a cyclic permutation of the colors.

Figure 5.13 shows the same pattern colored with three colors, but in a different way. This time H is a class-equivalent subgroup of G; it is also a normal subgroup. All of the elements of H map each colored set onto itself. The other translations of G permute the colors cyclically.

I cannot resist assigning some exercises here. For each of the colored patterns in Figures 5.14 and 5.15, determine:

    the index of H;

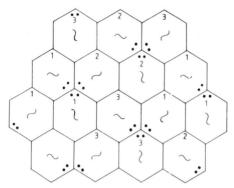

Figure 5.13. Another three-coloring of the pattern of marked hexagons.

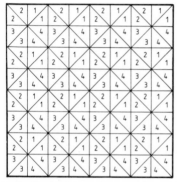

Figure 5.14. Find H and determine its properties.

whether H is a class-equivalent or a translation-equivalent subgroup (or neither);

whether H is a normal subgroup.

And, when H is not a normal subgroup, find another coloring. Uncolored patterns are provided for this purpose in Figures 5.16 and 5.17. Solutions can be found in Appendix 2.

5.4 Color groups, color symmetry, and colorings

We have seen that each subgroup of a symmetry group G determines a one-to-one correspondence between the elements of G and

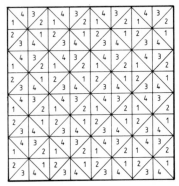

Figure 5.15. Find H and determine its properties.

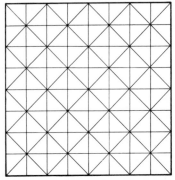

Figure 5.16. Is there another coloring of this pattern for the subgroup H of Figure 5.14?

a group of color permutations. Denoting this permutation group by S, we can, if we wish, denote this correspondence by a *pair* of groups, (G, S). If we know how multiplication in each of these groups corresponds to the other, we can form a new group of the ordered pairs. Such a group is called a *color group*.

The problem with color groups, and this ordered pair notation, is that they do not tell us very much. The same permutation group S can be associated with G via many different subgroups H; to know how a color group works we have to know what H is. It is, of course, an interesting problem to work out the H's that correspond to each choice of S.

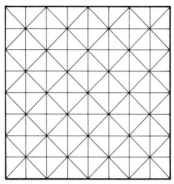

Figure 5.17. Is there another coloring of this pattern for the subgroup *H* of Figure 5.15?

Color symmetry is a broader term that encompasses the group theory of colored patterns that we have discussed here, but goes beyond it. The requirement of perfect colorings is very restrictive, perhaps too restrictive for some purposes. For example, some people argue that we should be allowed to color right cosets as well as left. For non-normal *H*'s, this will not associate each element of *G* with a unique color permutation, but perhaps that is a reasonable price to pay for the larger class of patterns we obtain this way.

Even more generally, one might consider colorings that are not particularly symmetrical, but which have some other regularity features. Such patterns cannot be described by groups, but there may be other, more appropriate ways to describe them. Escher's methods are of particular interest here.

The continuing usefulness of colored patterns in crystallography will depend on the significance of the features of the pattern that they illuminate. As the scope of crystallography continues to broaden, the class of colored patterns that are relevant to it may broaden also.

Chapter 6

Classification, and the *International Tables*

6.1 The classification problem

One can argue that all of pure science is, at heart, a project of classification and enumeration—the ancient problems of determining similarities and differences, discriminating between primary and secondary characteristics, delineating categories. It is a problem as old as natural philosophy, as new as the most recently discovered subatomic particle.

The distinction between science on the one hand and nature on the other becomes clear when we realize that any classification scheme is a human construction. When we classify things, we are placing them in categories that we ourselves have defined. The poet Robert Frost once wrote, 'Before I'd build a wall, I'd want to know what I was walling in and what I was walling out'. When we construct categories and place things in them, that is exactly what we are doing: deciding what to wall in and what to wall out.

The history of science shows us that scientific classifications have changed over time, depending on what were thought, in a given era, to be the salient, fundamental attributes of the objects being classified.

As we noted in Chapter 1, in ancient times Plato argued that matter was composed of four 'elements': earth, air, fire and water. Aristotle argued that these elements were not primary, but mixtures of four 'qualities': hot, cold, moist and dry (fire was a mixture of hot and dry, water of cold and moist, and so forth). It need hardly be noted that our concept of element is very different today, though we do recognize the liquid, solid, gas and plasma *states* of matter. (I am not sure of the present status of the 'qualities', but the names of some of today's subatomic particles are intriguing.)

For a more current example, we can turn to biology. In pre-DNA days, animals were classified by their morphological characteristics and lines of evolutionary descent. Today DNA patterns are thought to be the more fundamental biological characteristic. Do these classification schemes agree? If not, then the concept of 'species' will have to be reexamined (once again).

The classification of crystals has also changed over the centuries. In the Middle Ages (and again today for those who believe they live in a 'New Age'), classification was implicit but functional: crystals were grouped according to their talismanic and medicinal 'virtues'. The first person to make what we would call a modern classification was G Agricola, in 1530; his categories included color, taste and smell, but not symmetry. In the nineteenth century, the important characteristics of crystals were thought to be their external forms, the symmetry of their ideal forms, and their physical properties. In the twentieth century, after the discovery of x-ray diffraction, the fundamental characteristic of a crystal has been its atomic pattern, with emphasis on its space group symmetry. It is too soon to say what the classification scheme of the twenty-first century will be; it is not certain that symmetry will continue to be the central paradigm.

It is important to remember that symmetry is not now and, even in modern times, never has been the only way to classify crystals. Fedorov, who devoted years to enumerating the space groups, did not consider symmetry to be of primary importance because 'symmetry is a property of individual figures'. That is, it can be changed by an arbitrarily small deformation such as an infinitesimal elongation or shear. One well-known crystallographer told me of his plan, unfortunately never carried out, to deliver a lecture to an international congress with the title, 'To Hell With The Unit Cell!'. In his view, the network model of crystal structure, which emphasizes combinatorial over metric properties, is more fundamental than the space group.

6.2   Why symmetry?

Nevertheless, symmetry arguments have been increasingly effective in crystallography since the work of Haüy. We can distinguish two lines of development, geometric and algebraic, focused primarily on symmetry elements and symmetry groups, respectively.

As we saw in Chapter 2, the symmetry elements of a figure are the rotation axes, reflection planes, and so forth left fixed by its point symmetry operations (in the modern terminology of linear algebra, they are the eigenspaces of these linear maps). Haüy introduced the concept of an axis of symmetry in 1822 to explain regularity of form. For him, the symmetry of crystal form was axiomatic; he did not try to explain why the juxtaposition of identical blocks should result in a symmetrical stack.

In 1868 A Gadolin showed that crystals can be classified according to the configuration of their symmetry elements. In other words, to each crystal corresponds a three-dimensional 'symmetry diagram', or configuration of reflection planes and rotation axes (Figure 6.1). If the diagrams for two crystals are identical, then the crystals belong to the same class.

Gadolin's motivation for this work is of more than historical interest:

> *The system I am going to propose is only the result of a constant battle against the lack of precision in the ideas which I have been able to extract from works on crystallography on everything that has to do with the principle of classification.*

Gadolin's plaint was heeded. Symmetry theory, the ruling principle of crystallographic classification today, may be confusing and even controversial, but it can no longer be accused of lack of precision.

The early development of symmetry theory went hand in hand with the development of the mathematical theory of groups, and later it was considered an important example of the effectiveness of group theory in the study of geometry. Indeed, it was Klein, the principal evangelist of this approach to geometry, who suggested the problem of enumerating the space groups to Schoenflies. After 1912, the year in which x-ray diffraction was discovered, the success of space group theory as a tool in structure determination consolidated the preeminent role of symmetry in crystallography.

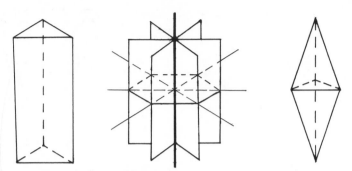

Figure 6.1. The symmetry elements of a triangular prism and a triangular bipyramid are the same.

6.3 Crystallographic classifications

Today crystals are classified by space group, by crystal class, by crystal system, by Bravais class, and in other ways as well. All of these categories are related to one another, and one task of this chapter is to explain both the categories and the relations.

The first thing to note is that all of these groups, systems and classes are abstract ideas; they are equivalence classes of mathematical structures. For example, when we assign a crystal to a space group we are really assigning it to a class of isomorphic groups.

From the point of view of group theory, two space groups, point groups, lattices, or more general orbits are equivalent if there is a still larger group under which they belong to the same orbit or orbit type (Figure 6.2), that is, if their stabilizers are conjugate with respect to this larger group.

The requirement of conjugacy partitions the set of all space groups, or the set of all lattices, or the set of all orbits, into classes. These classes are called the *strata* of the set. This provides a theoretical basis for most of the multitude of crystallographic classifications that are used in the *International Tables*.

How does it work in practice? Let us consider the classification of each of the crystallographic 'objects' mentioned above.

Point groups. The 32 crystallographic point groups are actually the strata, or conjugation classes, of the finite subgroups of $O(3)$

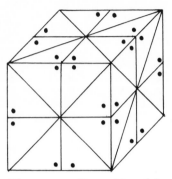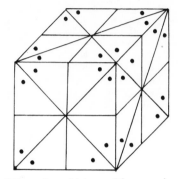

Figure 6.2. Two orbits of the group of the cube which are 'the same': both have trivial stabilizers.

that satisfy the crystallographic restriction. These are sometimes called *geometric crystal classes*, to distinguish them from the *arithmetic crystal classes*, which are a refinement of this classification. We have seen that in order to construct the space groups we need to know exactly how the point group acts on the lattice. This means we need to know how it transforms the generators of the lattice. A point group can be represented by a group of matrices, and since it acts on a lattice, the matrices can be written with integer entries. The entries in the matrix will change as we change the generators of the lattice; a change of generators leads to a matrix which is conjugate to the original one by an integral linear transformation. But two matrices representing the same symmetry operation (thus conjugate under an orthogonal transformation) need not be integrally conjugate. If they are not, then they act on lattices differently, and so we place them in different arithmetic classes.

For example,

$$\begin{pmatrix} 1 & 0 \\ 0 & -1 \end{pmatrix} \quad \text{and} \quad \begin{pmatrix} 0 & 1 \\ 1 & 0 \end{pmatrix}$$

both represent reflections in two-dimensional space but there is no integer matrix which conjugates one to the other. And, indeed, the matrix entries tell us that the transformations act differently on the lattices (Figure 6.3).

Lattices. We began the classification of lattices in Chapter 3.

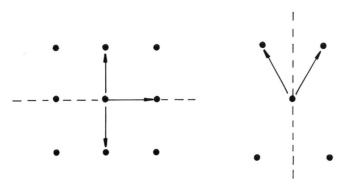

Figure 6.3. Two different actions of reflections on a lattice.

Let us briefly summarize what we found. Every lattice is an orbit of a translation group, and each translation group is generated by a set of linearly independent translations. Any triple (or, in n-dimensional space, n-tuple) of independent vectors can be transformed into any other by a linear map, and so, with respect to the group of all linear maps, all lattices are alike. Distinctions among lattices appear when we consider how the point groups act on them. We can partition the set of all lattices into classes, or strata, according to stabilizers of their origins. Those conjugation classes (in $O(3)$) of the point groups which serve as stabilizers of lattice origins are called the *crystal systems*. There are seven of them (Figure 3.3).

In the *International Tables*, however, the number of crystal systems is six (Figure 6.4); the trigonal and hexagonal lattices are placed in the same system. The number of systems is an old and continuing argument that goes back many years. (Could this be the source of the idiomatic expression of 'being at sixes and sevens'?) But, like centered cells, six crystal systems are not compatible with the requirements of a coherent mathematical theory: there are seven strata.

However many crystal classes one believes there should be, we still need to distinguish among different lattices in the same system, and there is no disagreement about the number of lattice classes, fourteen. These *Bravais classes* are the arithmetic classes of the crystal systems. (This is a modern interpretation of the criterion used by Bravais to distinguish the lattice types.) The Bravais

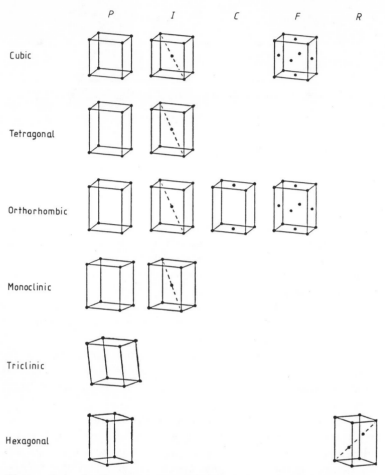

Figure 6.4. In the *International Tables* and much of the crystallographic literature, there are six crystal systems.

classes are distinguished notationally by the crystallographic symbols P, F, I and so forth (Figure 3.3). To the crystallographer, these symbols indicate whether or not the unit cell is 'centered', and if so, how. In any case these letters serve well as labels.

Thus lattices are classified in three ways, each a refinement of the preceding one, according to the strata to which they belong under the groups of all linear, orthogonal and integral transformations.

Space group orbits. Because any point of space can be chosen as the starting point for the creation of an orbit, any space group has infinitely many orbits. Again, we identify orbits which have conjugate stabilizers. For example, the two orbits of the plane group p4m shown in Figure 6.5 are essentially 'the same'.

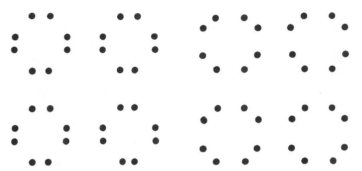

Figure 6.5. These two orbits of the plane group p4m are 'the same'.

Conjugation assigns space group orbits to strata, called *Wyckoff positions* in the *International Tables*. The group to which the strata belong can be larger than the space group itself, because the motion of superposition does not always belong to the space group. Some of the other Wyckoff positions for p4m are shown in Figure 6.6.

Figure 6.6. The Wyckoff positions for the plane group p4m (generic orbits are shown in Figure 6.5).

The space groups themselves. A single space group involves metric concepts: it consists of distance-preserving isometries, and its translation vectors have definite lengths. Although an x-ray crystallographer trying to determine the structure of a given crystal does need to determine the lengths of the edges of the unit cell, she will assign the crystal to a space group type in which such differences are ignored, as long as the symmetry remains the same.

The set of all space groups is partitioned into strata by the action of the group of all linear transformations and translations. There are 219 of these strata, which are also the isomorphism classes of the space groups. The number becomes 230 if one distinguishes between orientation-preserving and orientation-reversing conjugacy transformations.

Space groups are also assigned to *Laue classes*; the Laue classes are the possible point groups for x-ray diffraction patterns. The Laue class of a space group is its point group, to which inversion has been added if it was not already there (Figure 6.7).

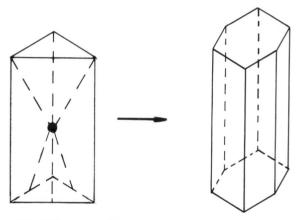

Figure 6.7. When we add inversion to the symmetry group of the triangular prism, we obtain the symmetry group of the hexagonal prism.

There are 11 Laue classes: exactly 11 of the 32 classes of crystallographic point groups contain inversion. Since the point group of a lattice always contains inversion, it is a Laue class. However, there are only seven crystal systems, not 11, because for four of the Laue classes, any lattice with that symmetry actually has more. In other words, Laue groups are not always maximal point groups for lattices.

6.4   Translating a page of the *International Tables*

We will study one example from the *Tables*, space group No 92 (Figures 6.8 and 6.9).

$P\,4_1\,2_1\,2$ D_4^4 4 2 2   Tetragonal

No. 92 $P\,4_1\,2_1\,2$ Patterson symmetry $P\,4/m\,m\,m$

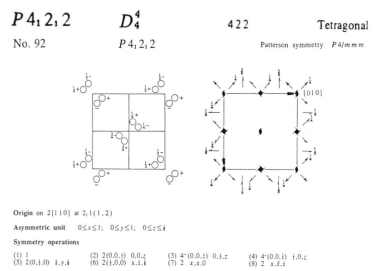

Origin on $2[1\,1\,0]$ at $2_1\,1\,(1,2)$

Asymmetric unit $0 \le x \le 1$; $0 \le y \le 1$; $0 \le z \le \frac{1}{8}$

Symmetry operations

(1) 1	(2) $2(0,0,\frac{1}{2})$ $0,0,z$	(3) $4^+(0,0,\frac{3}{4})$ $0,\frac{1}{2},z$	(4) $4^-(0,0,\frac{1}{4})$ $\frac{1}{2},0,z$
(5) $2(0,\frac{1}{2},0)$ $\frac{1}{4},y,\frac{1}{8}$	(6) $2(\frac{1}{2},0,0)$ $x,\frac{1}{4},\frac{1}{8}$	(7) 2 $x,x,0$	(8) 2 $x,\bar{x},\frac{1}{4}$

Figure 6.8. The first page of the *International Tables* description of the space group $P4_1 2_1 2$. Reproduced by permission of the International Union of Crystallography.

This group is denoted by the symbol $P4_1 2_1 2$. The first question is, what does this symbol mean? To understand this, we dissect it in the following steps.

First, separate the capital letter, in this case P, from the rest of the symbol: $P\quad 4_1 2_1 2$.

Next, suppress all the subscripts: the symbol becomes $P\quad 422$.

If there are any lower-case letters other than m, change them to m's (there are no other lower case letters in this example, but they occur in many space groups, for example $I4_1/m$ md).

What is left tells you the point group of the space group (the notation was explained in Chapter 2), whether the cell is centered, and if so, how. (P always means that the cell is not centered; it is a true unit cell.)

CONTINUED   No. 92 $P\,4_1\,2_1\,2$

Generators selected   (1); $t(1,0,0)$; $t(0,1,0)$; $t(0,0,1)$;   (2);   (3);   (5)

Positions

Multiplicity, Wyckoff letter, Site symmetry		Coordinates		Reflection conditions
				General:
8	b	1	(1) x,y,z (2) $\bar{x},\bar{y},z+\frac{1}{2}$ (3) $\bar{y}+\frac{1}{2},x+\frac{1}{2},z+\frac{1}{4}$ (4) $y+\frac{1}{2},\bar{x}+\frac{1}{2},z+\frac{3}{4}$ (5) $\bar{x}+\frac{1}{2},y+\frac{1}{2},\bar{z}+\frac{1}{4}$ (6) $x+\frac{1}{2},\bar{y}+\frac{1}{2},\bar{z}+\frac{3}{4}$ (7) $y,x,\bar{z}$ (8) $\bar{y},\bar{x},\bar{z}+\frac{1}{2}$	$00l:\ l=4n$ $h00:\ h=2n$

Special: as above, plus

4	a	..2	$x,x,0$ $\bar{x},\bar{x},\frac{1}{2}$ $\bar{x}+\frac{1}{2},x+\frac{1}{2},\frac{1}{4}$ $x+\frac{1}{2},\bar{x}+\frac{1}{2},\frac{3}{4}$

$0kl:\ l=2n+1$
or $2k+l=4n$

Symmetry of special projections

Along [001] $p\,4g\,m$
$a'=a$ $b'=b$
Origin at $0,\frac{1}{2},z$

Along [100] $p\,2g\,g$
$a'=b$ $b'=c$
Origin at $x,\frac{1}{4},\frac{1}{8}$

Along [110] $p\,2g\,m$
$a'=\frac{1}{2}(-a+b)$ $b'=c$
Origin at $x,x,0$

Maximal non-isomorphic subgroups

I $[2]P\,4_1\,1\,1\,(P\,4_1)$ 1; 2; 3; 4
     $[2]P\,2_1\,2_1\,1\,(P\,2_1\,2_1\,2_1)$ 1; 2; 5; 6
     $[2]P\,2_1\,1\,2\,(C\,2\,2\,2_1)$ 1; 2; 7; 8
IIa   none
IIb   none

Maximal isomorphic subgroups of lowest index

IIc $[3]P\,4_3\,2_1\,2\,(c'=3c)$; $[5]P\,4_1\,2_1\,2\,(c'=5c)$; $[9]P\,4_1\,2_1\,2\,(a'=3a,b'=3b)$

Minimal non-isomorphic supergroups

I $[3]P\,4_1\,3\,2$
II $[2]I\,4_1\,2\,2$; $[2]C\,4_1\,2\,2\,(P\,4_1\,2\,2)$; $[2]P\,4_2\,2_1\,2\,(2c'=c)$

Figure 6.9. The second page of the *International Tables* description of the space group $P\,4_1\,2_1\,2$. Reproduced by permission of the International Union of Crystallography.

The small letters other than m indicate glide reflections, and the subscripts indicate screw rotations—these correspond to the nontrivial solutions of the Frobenius equation (Chapter 4). If there are no such symbols, you know that the group is symmorphic.

Now that the space group symbol has been deciphered, we can proceed from left to right across the top of the page. Here we find, first, the symbol for the point group that Schoenflies introduced and is still used in chemistry; next, the international symbol for the point group, and finally, the crystal system to which the space group is assigned.

The next line gives the space group number (this is of importance only for the interior consistency of the *Tables*), the 'long' space group symbol which spells out in more detail exactly which mirror and glide planes are orthogonal to which rotations axes, and a group called the Patterson Symmetry (it is the symmetry group of the Patterson function used by x-ray crystallographers). This group is necessarily symmorphic, and its point group may be larger than the point group of the space group.

Moving down the page, we come to two diagrams, one of which is a unit cell's worth of a generic orbit for the space group, on which heights above and below the plane of the page are indicated; the other is a unit cell's worth of its symmetry element diagram. Sometimes, neither of these diagrams makes much sense as it appears; it is necessary to draw several adjacent copies of each to understand what the pattern is really like (Figure 6.10).

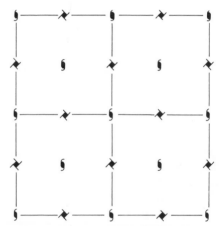

Figure 6.10. With several copies of the drawings in the table, we see the symmetry more clearly.

It is helpful to draw these diagrams on tracing paper or, better yet, acetate sheets, so that they can be superimposed. If you spend a little time studying the patterns, you will begin to understand how the group works. It is sometimes helpful to try grouping the points together in various ways. For example, I found that this orbit of $P4_12_12$ made more sense to me when I grouped the points into sets of four, and envisaged each set as the vertices of a rectangle (Figure 6.11).

Next, below the diagrams, the choice of origin used for the diagrams and the computation of symmetry operations is given; this

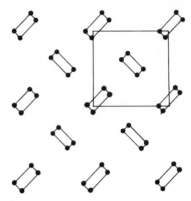

Figure 6.11. This grouping of points is helpful in understanding the pattern of Figure 6.8.

choice of course is arbitrary. In the *International Tables* it is chosen to lie at a point which is fixed by some symmetry operation, if that is possible; in our example it was chosen to lie on a two-fold axis. This means, however, that it does not lie on a four-fold axis, which in some ways might seem to be a more natural choice.

Below the origin statement is a description of the 'asymmetric unit', that portion of the unit cell of the lattice which can serve as a motif for a generic orbit of the group. It is a 'smallest part' in the sense of Chapter 2.

Next, the symmetry operations (t, p) are given for each symmetry operation of the point group, followed by a description, in terms of coordinates, of the corresponding symmetry element. Each is assigned a number, for further reference.

Now we move on to the next page. This page begins with a list of space group generators: three translations and an appropriate number of operations from the preceding list, given only by the numbers that had been assigned to them.

Under 'positions', we find a description of the orbit classes (Wyckoff positions) of the space group, beginning with the generic one illustrated on the preceding page. The number of points of the orbit in a single unit cell, a letter denoting the orbit class, which need not concern us here, and a symbol indicating the stabilizer subgroup and its location relative to the symmetry element diagram precede the description, in coordinates, of the position of the orbits themselves.

On the far right-hand side of the page are listed 'reflection conditions'. These indicate the directions in which the points of the orbit lie on planes which are more closely spaced than the lattice structure would indicate; this can happen when glide planes and axes of screw rotation are present. For example, in the orbits of $P4_12_12$, points lie on four equally spaced planes within the unit cell, orthogonal to the direction of the four-fold axis, because of the 4_1 axis. This means that the spacing between these planes is one quarter the spacing between lattice planes (Figure 6.12). In the x-ray diffraction pattern the dual lattice spacing will appear to be four times as large as it actually is. Thus the dual lattice spacing will occur in multiples of four. (The use of centered cells also causes this apparent anomaly.)

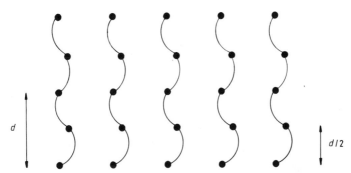

Figure 6.12. Screw axes decrease the interplanar distance.

Interestingly, this is exactly the same phenomenon that causes some of the anomalies in the application of Bravais' morphological law. This was pointed out by Donnay and Harker in 1937; their success in resolving some of these anomalies helped to convince crystallographers that space groups should be taken seriously.

Next on this page is a list of the plane groups that result from the projection of the generic space group orbit along certain lattice lines.

Finally, we come to a list of the maximal subgroups of the space group, and of the groups of which the space group itself is a maximal subgroup (the minimal supergroups). The maximal subgroups of a group are those which lie on the top rung of its subgroup diagram. As we have seen, there are two types of maximal subgroups:

those whose translation subgroups are identical with that of the space group, and those whose point groups are identical with that of the space group. In the *Tables* they are called types I and II, respectively.

These categories are divided into subcategories in the *Tables*, but this is a good place to stop.

Chapter 7

N-dimensional Crystallography?

*While these starlets are falling, they consist of three feath-
ered diameters, joined crosswise at one point, with their six
extremities equally distributed in a sphere; consequently, they
fall on only three of the feathered prongs, and tower aloft with
the remaining three, opposite those on which they fall, on the
same diameters prolonged, until those, on which they rested,
buckle, and the remainder, until then upright, sag onto the
level with the former in the gaps between them.—J Kepler,
The Six-Cornered Snowflake, 1611.*

7.1 The view from N-dimensional space

Kepler was surely not the first person to wonder why snowflakes
have branched hexagonal forms, but it is likely that he posed the
most imaginative solution. Perhaps the forms that we see are not
the original ones, he suggested. Perhaps, as it falls from the sky, the
snowflake is three dimensional, with its branches pointing to the
vertices of a regular octahedron; then, when it lands, it collapses
to a planar hexagonal form (Figure 7.1).

Kepler immediately rejected this fanciful hypothesis and, as we
saw in Chapter 1, went on to more fruitful musings on snowflake
structure. But the idea of interpreting two- and three-dimensional
forms as collapsed higher-dimensional ones still intrigues us today.
Can you see, in Figure 7.2, a collapsed five-dimensional cube?

Figure 7.1. Kepler's explanation of the external form of the snowflake.

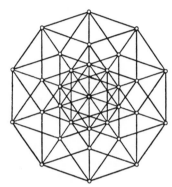

Figure 7.2. A projection of a five-dimensional cube onto a plane.

Today mathematical crystallography is confronted with the task of identifying and classifying spatial patterns which, though (r, R) systems, are not regular ones. They are not orbits of groups, yet they have sufficient long-range order to produce nice diffraction patterns. That is, they produce diffraction patterns with bright spots whose symmetry tells us that the pattern that produced them does not have a lattice structure. Although at first sight they seem disordered, many of these patterns can be interpreted as projections of sets of points which belong to regular (r, R) systems in some higher-dimensional space.

Of course, no one seriously suggests that the crystals we try to model by nonperiodic (r, R) systems of this type really fall into our world from some higher-dimensional space. The justifications for studying such models are that they help us to discover some of the order properties that are possible in nonperiodic patterns, and that they facilitate computation. (In particular, it is not difficult to compute the diffraction patterns of projected sets, since one can use standard techniques in higher dimensions.)

It is a straightforward, if sometimes tedious, task to extend the concepts discussed in earlier chapters to higher dimensions. The point groups in N dimensions are the finite subgroups of the group $O(N)$ of symmetries of the N-dimensional sphere. As in lower dimensions, N-dimensional lattices are orbits of discrete groups generated by N independent translations, and they are classified by their stabilizers. There is an analogue of the crystallographic restriction (Theorem 1.2) in each dimension. The N-dimensional space groups are solutions of the Frobenius equation. There are complete tables of the space groups for $N = 4$, and various committees are considering the feasibility of constructing partial tables for $N = 5$ and $N = 6$.

In fact, N-dimensional crystallography was a subject of mathematical research before three-dimensional lattices and space groups were accepted by crystallographers! At the turn of the century, Voronoï and Minkowski studied the geometry of N-dimensional lattices; their work must have inspired Hilbert to include, as another part of his 18th problem, the question whether the number of crystallographic groups in N-dimensional space was finite, as it was by then known to be for $N = 3$. In several papers written between 1910 and 1912, Bieberbach showed that the answer to this question was yes. In that same period, Frobenius published his equation, which is valid in any dimension. Thus by the time that x-ray diffraction was discovered, the theory of N-dimensional space groups was already well developed.

There is no particular reason why the symmetry group of a crystal must have the same dimension as that of the crystal. The more important question is whether the N-dimensional point of view is sufficiently rich to afford a means of classifying nonperiodic crystals, or whether it is too restrictive and artificial.

7.2 Projections

Projections are not new to crystallography; as we have seen, the *International Tables* include projections of three-dimensional structures onto planes of high symmetry. These planes are lattice planes, and the three-dimensional structures can be reconstructed by stacking them. In other words, it is understood that each point in the projection represents infinitely many points. This is easier to visualize if we look at the analogous situation one dimension lower, the projection of a plane lattice onto a line. Figure 7.3 shows that when a two-dimensional lattice is projected orthogonally onto a line l which contains two (and hence infinitely many) lattice points, then all the lattice points on each line orthogonal to l project to the same point. The projected set is always discrete.

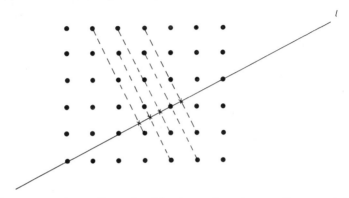

Figure 7.3. A two-dimensional lattice projected onto a line containing lattice points is a discrete set.

When a three-dimensional lattice is projected onto a lattice plane, the projected set is periodic. To use projections to interpret nonperiodic point sets, we need to project onto 'irrational' planes (planes with at most one lattice point). But then the lattice projects to a dense set of points (Figure 7.4).

More generally, an N-dimensional lattice projected onto an irrational subspace of any dimension is not an (r, R) system because it is not discrete, and so is not a suitable model for any discrete physical system. However, you can have your cake and eat it too if you project only part of the lattice, a part lying in some bounded strip (Figure 7.5). Then the projected set is both discrete and nonperiodic—and is a very interesting (r, R) system.

The diffraction patterns produced by such projected (r, R) systems are not quite like those produced by regular ones: although we see bright spots against a dark background, the spots are actually dense. However, some are so much brighter than their neighbors, and others so much fainter, that we only see the bright ones.

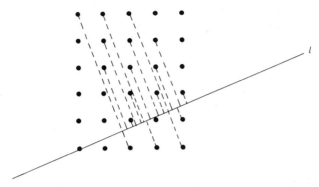

Figure 7.4. If l contains only one lattice point, then each point of the lattice projects to a different point on the line and we get a dense set.

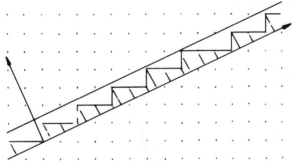

Figure 7.5. When the points in the strip are projected to the line, we get a nonperiodic (r, R) system.

P M de Wolff introduced projections from higher dimensions into crystallography in the early 1970's, showing that one can use regular systems of points in four-dimensional space to explain the diffraction patterns of 'modulated' crystals. The atoms in these crystals are arranged periodically, but the constituent lattices of the patterns are not congruent—they may not even be commensurate. De Wolff and his colleagues found that the three-dimensional patterns they were studying could be obtained by projection; this enabled them to classify the diffraction patterns in an elegant and useful way.

The modulated crystals studied by de Wolff have crystallographic symmetry. But, as we will see, diffraction patterns of projected point sets may also have noncrystallographic symmetry. First, for historical, pedagogical and aesthetic reasons, we take a detour into the intriguing topic of Penrose tiles, and show why they are of interest for crystallography.

7.3   The Penrose tiles

In 1981, A L Mackay exhibited an optical transform of the point set in Figure 7.6(a); the diffraction pattern, with its five-fold symmetry (Figure 7.6(b)), dramatically demonstrated the nonperiodic nature of the pattern of points.

This point set is neither a lattice nor a union of lattices. Then what kind of order does it have? Although the points at first appear to be randomly scattered, they must have some hidden long-range order; otherwise they could not produce such a nice diffraction pattern.

We might investigate the pattern by constructing the Voronoï cells of the points, but in this case the Voronoï tiling does not shed much light on the matter (Figure 7.7). But if we link the points as shown in Figure 7.8, we see a remarkable pattern. The points are the vertices of a tiling of the plane by two kinds of rhombi.

The pattern in Figure 7.8 is part of one of the famous tilings discovered by R Penrose. The rhombi, one thin and one thick, have equal edge lengths and small angles of 36° and 72°, respectively. Thus ten thin or five thick rhombi can be fitted around a point. Remarkably, the tiling can be continued to cover the plane nonperiodically.

The symmetry of these tilings is very interesting. One finds finite regions of five-fold symmetry almost everywhere, but the symmetry breaks down as the size of the region increases. It is interesting that Kepler observed the same thing in 1619, in his study of the tilings of the plane by polygons (Figure 1.15). We reproduce here his diagram Aa (Figure 7.9).

The Penrose rhombi are reminiscent of unit cells, but there are two shapes instead of one. Like unit cells, these rhombi can form periodic tilings (Figure 7.10), but these are of little interest.

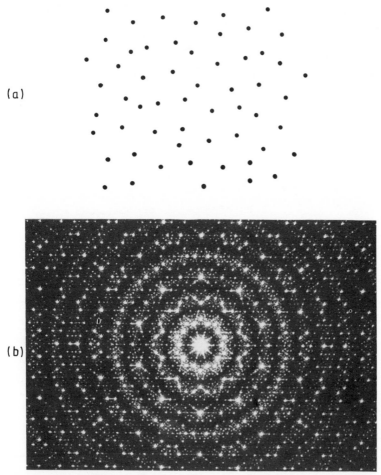

Figure 7.6. (a) This point set produces the diffraction pattern shown in (b). Optical diffraction pattern made by Dr G Harburn (Cardiff) for A L Mackay.

To outlaw periodic arrangements, we need some sort of matching rules that force the tiling to be nonperiodic. As Penrose showed, such rules exist; they can be specified by placing arrows (of two kinds or colors) on the edges of the rhombi. According to these rules, rhombi can only be juxtaposed if the adjacent arrows are of the same type and point the same way (Figure 7.11). If the rules

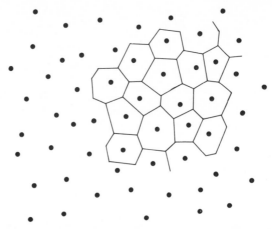

Figure 7.7. The Voronoï cells of the points in Figure 7.6(a).

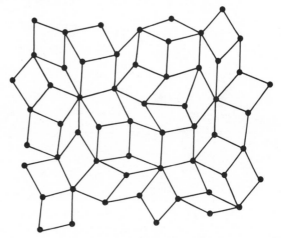

Figure 7.8. Another tiling interpretation of the same point set.

are followed, then the tiling will always be nonperiodic.

Although the Penrose tilings are nonperiodic, they have a high degree of order. For example: they are self-similar in the sense that they contain rescaled copies of themselves; any finite region of each patterns occurs infinitely many times, not only in that tiling but in all the others that can be made with the same tiles (and rules);

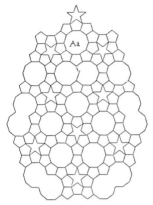

Figure 7.9. Kepler's experiment with five-fold symmetry in the plane.

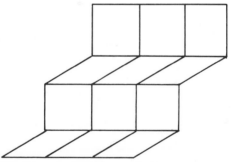

Figure 7.10. The Penrose rhombi can tile the plane periodically if no restrictions are placed on the way we put them together.

they have arbitrarily large finite regions with five-fold symmetry.

7.4 De Bruijn's interpretation

The connection between the Penrose tiles and higher-dimensional lattices was discovered by N G de Bruijn in 1981. His work has become the basis for all recent work on the projection model in

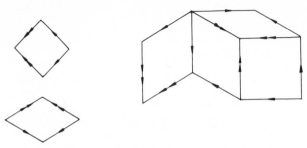

Figure 7.11. Matching rules for the Penrose tiles.

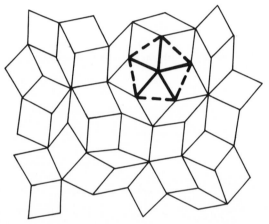

Figure 7.12. The edges of the tiles in a Penrose tiling have only five different orientations.

crystallography.

De Bruijn first observed that although the tiles seem to meander every which way, in fact their edges have only five different orientations: they all point from the center to the vertices of a fixed regular pentagon (Figure 7.12).

He then noted that because the tiles are parallelograms, any edge defines an infinite band of tiles (Figure 7.13), just as the centrosymmetric faces of a parallelohedron lie in circular bands. In the Penrose tilings, these bands run in five directions, each direction being orthogonal to a set of parallel edges. Bands of each direction reappear over and over throughout the tiling. They are not

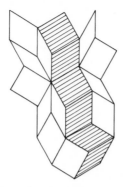

Figure 7.13. Any edge defines an infinite band of tiles.

regularly spaced, but they are not arbitrarily far apart. One can represent each band by a straight line, and the corresponding set of 'parallel' bands by a 'grid': a *grid* is a series of parallel straight lines together with a unit vector orthogonal to them (Figure 7.14). For our purposes we can assume that the lines of a grid are equally spaced.

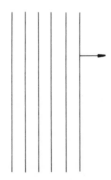

Figure 7.14. Each set of parallel bands of rhombi in a Penrose tiling can be represented by a grid.

Now consider five identical grids, superimposed so that their unit vectors point to the vertices of a pentagon, but shifted so that no more than two lines cross at any point (Figure 7.15(a)). De Bruijn called such a configuration a *regular pentagrid*. Using it, he was able to reconstruct the Penrose tilings (complete with

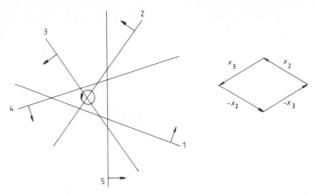

Figure 7.15. Every crossing point in a regular pentagrid (a) corresponds to a tile (b).

matching rules). His key idea was to associate a tile to every point of intersection of the lines of the pentagrid, by placing end to end the unit vectors orthogonal to the grids that one crosses by encircling that intersection point in a simple loop (Figure 7.15(b)).

As we move along a single line of one of the grids, we encounter an infinite sequence of crossing points. The tiles constructed for each of them constitute a band of the tiling (Figure 7.16). In this way we can fill the plane with Penrose tiles. De Bruijn proved that every possible Penrose tiling can be constructed from grids. (Some Penrose tilings correspond to nonregular pentagrids; these, he showed, are limiting cases.)

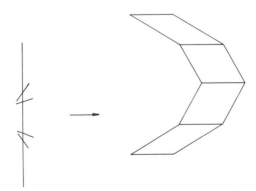

Figure 7.16. The pentagrid determines the tiling completely.

There is a duality between the pentagrids and Penrose tilings: the nodes and cells of the pentagrids correspond to cells and nodes of the tilings, and de Bruijn used this to reconstruct the matching rules. Every mesh of the pentagrid can be assigned a location symbol, a pentuple of integers $(k_1, \ldots, k_5)$ specifying its location relative to the sequence of lines of each grid (first line of grid 1, eighth line of grid 2, and so forth). Since each mesh of the pentagrid corresponds to a vertex of the tiling constructed from it, each vertex is assigned in this way a set of five integer coordinates. It turns out that the sum of the coordinates, $k_1 + \ldots + k_5$, of each vertex of a Penrose tiling is always greater than 0 but less than 5; this sum is called the *index* of the vertex. The pattern of indices can be interpreted as rules for matching edges (Figure 7.17).

Now for the connection with higher dimensions. Since each vertex has five integer coordinates, we can identify it with a point of an integer lattice in five-dimensional space! Then the vertices of the Penrose tiling can be obtained by projecting these points onto a plane. This plane is not a lattice plane for the five-dimensional lattice, since the pattern is not periodic. But the projected set is an (r, R) system. Thus not all of the points of the lattice are projected; there must be some selection principle.

Indeed, the index of a vertex is the scalar product of the vectors $(k_1, \ldots, k_5)$ and $(1,1,1,1,1)$; the latter vector is the body diagonal of the unit cube of the lattice. Since the scalar product is proportional to the projection of the first of these vectors onto the second, the lattice points which project to vertices lie in the region that projects onto a finite segment of the line containing $(1,1,1,1,1)$. (But not all points in this region correspond to vertices; there are additional conditions.)

7.5 Generalized crystallography

The Penrose tilings have five-fold symmetry in finite regions; their diffraction patterns indicate that there is some sense in which the five-fold symmetry is a global property. At present, there is still no theory that explains the diffraction pattern in terms of the *local* properties of the tilings. But the projection model does provide a simple interpretation.

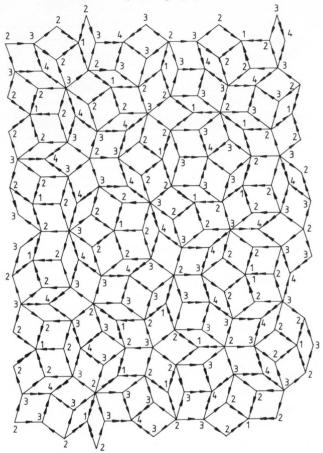

Figure 7.17. Interpreting the pattern of indices as matching rules. Reproduced by permission of The Royal Netherlands Academy of Arts and Sciences.

In Chapter 2 we noted that every isometry in three-dimensional space can be represented by a diagonal matrix with entries ±1 or by a matrix of the form

$$\begin{pmatrix} \cos\theta & -\sin\theta & 0 \\ \sin\theta & \cos\theta & 0 \\ 0 & 0 & \pm1 \end{pmatrix}.$$

The 2 × 2 block in this matrix represents a rotation in a plane.

Isometries in N dimensions can also be represented by matrices of the same general type: their diagonal entries are either ± 1, or 2×2 blocks representing rotations of two-dimensional planes, possibly through different angles. For example, in five-dimensional space a five-fold rotation about a line (such as the body diagonal of a hypercube) can be written in the form

$$\begin{pmatrix} \cos 2\pi/5 & -\sin 2\pi/5 & 0 & 0 & 0 \\ \sin 2\pi/5 & \cos 2\pi/5 & 0 & 0 & 0 \\ 0 & 0 & \cos 4\pi/5 & -\sin 4\pi/5 & 0 \\ 0 & 0 & \sin 4\pi/5 & \cos 4\pi/5 & 0 \\ 0 & 0 & 0 & 0 & 1 \end{pmatrix}.$$

This rotation rotates *two* planes simultaneously! The planes are orthogonal to the rotation axis and to one another. Each is *invariant* under the rotation: it is rotated onto itself.

Now consider the ordinary (primitive, integer) cubic lattice in five dimensions. The unit cell is a hypercube whose body diagonal is the line segment between $(0,0,0,0,0)$ and $(1,1,1,1,1)$. As we have seen, a five-fold rotation about this line as axis has two invariant planes. *Both planes are irrational: neither contains any lattice point except the origin!* If we project the hypercube onto one of them we obtain a two-dimensional configuration which is invariant under this higher-dimensional rotation (see Figure 7.2). If we project the bounded region of the lattice that we discussed in the last section, we obtain a nonperiodic (r, R) system. In fact, it is the set of vertices of a Penrose tiling.

Mackay's diffraction pattern can be computed from this model. The theory outlined in Chapter 3 applies in five (and all other) dimensions; then we use mathematical techniques to restrict the computations to the appropriate regions, and to obtain the diffraction pattern of the projected set.

In exactly the same way we can, by projection from six- to three-dimensional space, obtain a set of points which are the vertices of two kinds of rhombohedra, one thin and one thick. They are three-dimensional analogues of the Penrose tiles, and their diffraction patterns display icosahedral symmetry.

Evidently, we can start with this (or any other) lattice in N-dimensional space, find the invariant planes (or, more generally, subspaces) of the lattice rotations, and project bounded regions onto them. When the subspaces are irrational we will get candidates for generalized crystals: they will always have nice diffraction patterns. It is not always possible to interpret such sets as vertices of tilings, but in every case they are (r, R) systems with order properties very like those of the Penrose tiles.

Theoretically, the diffraction patterns of (r, R) systems produced in this way can have rotational symmetry of any order: the orders of rotation permitted by the crystallographic restriction increase with increasing N. This is one of the ways in which mathematical crystallography is 'enlarging its definition of homogeneity'. Whether real crystals are following suit in similar ways is of course another matter. Diffraction patterns with eight-fold, ten-fold, and twelve-fold symmetry have been found in quasicrystal research, but no other noncrystallographic symmetries have been observed.

At the present time, none of the (real) quasicrystal structures has yet been solved, and the validity of the Penrose tiling model is considered highly dubious (more general projection models fit the data better). The tilings are still useful, however, in helping us think about what geometrical properties generalized crystals might have. For example, although we can create tilings analogous to the Penrose tilings by projection from any dimension, it seems to be impossible to find matching rules for most of them. (Matching rules do exist in a very few cases, including both the two- and three-dimensional Penrose tiles, but it seems likely that in most cases no such rules exist.) Matching rules for tiles, or something analogous for more general point sets, would seem to be a useful feature of any model of an ordered but nonperiodic structure, since they can be thought of as a geometrical analogue of the chemical bond and thus may help us to understand how the structure grows. On the other hand, maybe it is possible to construct a theory of nonperiodicity with local rules that are somewhat more general.

The move to N dimensions, and then back to three by projection, is currently a very popular technique, but it is not clear yet that this is the best way to understand generalized crystals. It is rather artificial, and one would prefer a local model, one that gives some insight into growth processes. Nor is it clear that an N-dimensional interpretation is always possible: there are interesting nonperiodic (r, R) systems which do not appear to fit this model. But the study of this model broadens the scope of our understanding of pattern in an important way.

Understanding the relation between local and global order is the preeminent challenge for mathematical crystallography today; it is the newest chapter in the problem of structure and form. Many different mathematical tools will be needed to resolve this problem; increasingly, the emphasis will be on their interaction. Instead of separating the study of spatial patterns from the study of diffraction patterns, the intimate relation between the two will be exploited. Group theory, geometry, graph theory, number theory, tiling theory, Fourier analysis and techniques yet to be developed will all be needed to understand the fascinating new patterns we are discovering in the world of crystals.

Appendix 1

Cast of Characters

Our contemporary views on mathematical crystallography have been influenced by the natural philosophers and scientists of many eras and countries, writing in many languages. Those mentioned in this book are listed below.

Agricola, Georgius (1494–1555), German physician, and metallurgist, author of works on mining, geology, and mineralogy.

Aristotle (384–322 BC), Greek philosopher. His comments on Plato's theory of matter appear in *De Caelo* III, 306b.

Bieberbach, Ludwig (1886–1982), German mathematician. He proved that the number of space groups is finite in any dimension.

Bragg, William Henry (1862–1942) and **Bragg, William Lawrence** (1890–1971), British physicists. Father and son were pioneers in the development of x-ray diffraction crystallography.

Bravais, August (1811–1863), French naval officer and scientist of broad interests. He enumerated correctly the 14 types of three-dimensional lattices and studied the implications of lattice theory for crystal structure and properties.

Bruijn, Nicholas G de (1918–), Dutch mathematician. He discovered the connection between the Penrose tiles and lattices in five-dimensional space.

Burckhardt, J J (1903–), Swiss mathematician and historian of crystallography. By giving the space groups a modern mathematical interpretation, he helped to create the field of mathematical crystallography.

Curie, Pierre (1859–1906), French physicist, discoverer of piezoelectricity. His 'Principle of Symmetry' is widely quoted.

Delone, Boris Nicolaevitch (1890–1980), Russian mathematician. Together with colleagues and students, Delone simplified and extended the work of Fedorov and reformulated much of mathematical crystallography in terms of (r, R) systems.

Escher, Maurits Cornelius (1898–1971), Dutch graphic artist. His colored tessellations of the plane inspired research in mathematical crystallography.

Euler, Leonhard (1707–1783), Swiss mathematician of extraordinary breadth and depth. Euler's work on the combinatorial properties of polyhedra is a basic tool in the study of network models of crystal structure.

Fedorov, Evgraf Stepanovich (1853–1919), Russian crystallographer. One of the enumerators of the space groups, he also made many contributions to the tiling model of crystal structure.

Frankenheim, Moritz (1801–1869), German crystallographer. Frankenheim was the first to enumerate the 32 crystal classes. He was also the first to enumerate the three-dimensional lattices, but his list contained a duplication, later corrected by Bravais.

Frobenius, Georg (1849–1917), German mathematician, known for his work in group theory. His famous congruence is the basis for an algorithm for the construction of the space groups.

Gadolin, Axel (1828–1892), a Finn who was a Russian artillery professor and amateur crystallographer. He laid the basis for the classification of crystals by symmetry.

Haüy, René-Just (1743–1822), French abbé and mineralogist. His building-block theory of crystal structure led directly to the lattice model.

Hermann, Carl (1898–1961), German crystallographer. He made important contributions to the study of N-dimensional crystallography and the theory of subgroups of crystallographic groups.

Hilbert, David (1862–1943), German mathematician. One of the giants of twentieth century mathematics, his 1900 list of 23 problems shaped the course of subsequent research; the 18th concerned tilings, sphere-packings, and N-dimensional space groups.

Hooke, Robert (1635–1703), British scientist, member of the Royal Society. His monograph *Micrographia* contains detailed sketches of his observations with the microscope.

Jordan, Camille (1838–1922), French mathematician. His study of 'groups of motions' led directly to the development of the theory of space groups.

Kepler, Johann (1571–1630), astronomer and mathematican. Many of today's crystallographic concepts can be traced to his 1611 musings on the snowflake. He also contributed to tiling theory.

Klein, Felix (1849–1925), German mathematician. His advocacy of group theory as a tool in the study of geometry influenced the development of symmetry theory.

Laue, Max von (1879–1960), German physicist. His discovery in 1912 of the diffraction of x-rays by crystals unlocked the solid state.

Linnaeus, Carolus (1707–1778), Swedish naturalist best known for his classification of plants. His classification of crystals was less successful, but he was the first to emphasize the importance of shape.

Mackay, Alan L (1926–), British crystallographer, advocate of a broader scope for theoretical crystallography. He was the first to demonstrate that nonperiodic point sets could produce crystal-like diffraction patterns with 'forbidden' symmetries.

Pasteur, Louis (1822–1895), French scientist known for his work in crystallography, chemistry, microbiology and immunology. Early in his career he discovered the relation between optical activity and molecular asymmetry, and showed that crystal faceting can give clues to their presence.

Penrose, Roger (1931–), British physicist, inventor of the Penrose tiles.

Plato (427–348/347 BC), Greek philosopher. His theory of the structure of matter is discussed in his dialogoue *Timaeus*.

Schoenflies, Arthur (1853–1928), German mathematician. He was one of the enumerators of the space groups.

Sohncke, Leonard (1842–1897), German crystallographer. He enumerated the space groups containing only translations and rotations.

Steno, Nicolaus (1638–1686), Danish physician and scientist, later a priest. He argued that crystals were formed by the accretion of congruent units.

Voronoï, George (1868–1908), Russian mathematician. He is known for his work in number theory and the theory of tiling of N-dimensional space.

Wulff, George (1863–1925), Russian crystallographer. He is known in crystal growth theory for his construction of the ideal 'equilibrium form'.

Zassenhaus, Hans (1912–), American mathematician. He developed the algorithm used to enumerate the four-dimensional space groups.

Appendix 2

Answers to Exercises

The reader will have noticed that there are really only two official exercises in this book. In Chapter 1, you were asked to decide which of Kepler's tilings (Figure 1.15) can be extended to cover the whole plane. The 11 are shown as Exhibit A below. In Chapter 5 you were asked some questions about the subgroups of p4m illustrated in Figures 5.14 and 5.15. The subgroup in Figure 5.14 is of index four, is translation-equivalent, and is not a normal subgroup. Another coloring is shown in Exhibit B below. The subgroup in Figure 5.15 is also a translation-equivalent subgroup of index four but it is a normal subgroup, so there are no other colorings.

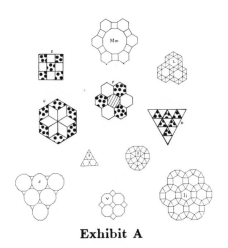

Exhibit A

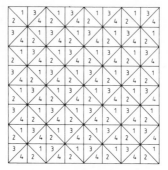

Exhibit B

Appendix 3

The finite subgroups of $O(3)$

The finite subgroups of the symmetry group of the sphere $O(3)$ are listed below. They are all derived from the equation of Theorem 2.6. In part A, my symbol for the group appears in the left-hand column and a polyhedron with that symmetry group is listed on the right. Since my notation differs slightly from that used by the International Union for Crystallography (in particular, I use rotary rotation and the IUCr uses rotary reflection), the two notations are compared in part B.

Part A

G	Polyhedron
nn	n-gonal pyramid
$n22$	n-gonal prism or antiprism
332	regular tetrahedron
432	cube
532	regular icosahedron
$nm\ nm$	n-gonal pyramid
n/m	—
$\widetilde{2n}$	scalene tetrahedron, when $n=2$
$n\mathrm{m}\ 2\mathrm{m}\ 2\mathrm{m}$	n-gonal prism
$n\mathrm{m}\ 2\ 2$	n-gonal antiprism
$3\mathrm{m}\ 3\mathrm{m}\ 2\mathrm{m}$	regular tetrahedron
$3\ 3\ 2\mathrm{m}$	pyritohedron
$4\mathrm{m}\ 3\mathrm{m}\ 2\mathrm{m}$	cube
$5\mathrm{m}\ 3\mathrm{m}\ 2\mathrm{m}$	regular icosahedron

Part B

H	IUCr
nn	n
$n22$	$n22$
332	23
432	432
532	—
nm nm	$n$mm
$n/$m	$n/$m
$\tilde{n}$	$\overline{2n}$
	$\overline{(n/2)}$
	$\bar{n}$
nm 2m 2m	$n/$m mm
	$\overline{2n}$m2
nm 2 2	$\overline{2n}$2m
	$\bar{n}$m
3m 3m 2m	$\bar{4}$3m
3 3 2m	m3
4m 3m 2m	m3m 2/m
5m 3m 2m	—

Further Reading

The literature of mathematical crystallography is very vast, spans centuries, and is written in many languages. There is a complete bibliography (up to 1978), prepared by the late W Nowacki and available from the International Union of Crystallography. My aim here is much more modest: to list the books and articles that I have found especially helpful in preparing the material for this book.

Historical works

Original works

Agricola, G, *De Natura Fossilium*, Basel, 1546 and 1558.

Bieberbach, L, 'Über die Bewegungsgruppen der Euklidische Räume', *Math. Ann.* **72** (1912) 400–12.

Bravais, August, 'Memoire sur les systèmes formés par des points distribués regulierement sur un plan ou dans l'espace', *Journal de l'Ecole Polytechnique* **XIX** (1850) 1–128; 'Etudes Cristallographiques', *Journal de l'Ecole Polytechnique* **XX** (1851) 101–276.

Bruijn, N G de, 'Algebraic theory of Penrose's nonperiodic tilings of the plane, I, II', *Nederl. Akad. Wetensch. Indag. Math.* **43** (1981) 39–52, 53–66.

Frobenius, G, 'Über die unserlegbaren diskreten Bewegungsgruppen', *Sitz. Konigl. Preuss. Akad. Wissenschaften*, 654–66, Berlin, 1911.

Gadolin, A, 'Deduction of all crystallographic systems and their subdivisions by means of a single general principle', *Annals of the Mineralogical Society*, St Petersburg, 1868 (in Russian).

Haüy, R-J, *Traité de Crystallographie*, three volumes, Paris, 1822.

Hooke, R, *Micrographia*, London, 1665.

Jordan, C, 'Memoire sur les groupes de mouvements', *Annali di matematica pura ed applicata* **II**, Ser. II. (1868) 167–215, 322–45.

Kepler, J, 'Strena seu de nive sexangula', *Francofurti ad Moenum*, 1611; *Harmonices Mundi*, 1619.

Linnaeus, C, *Systema Naturae*, Holmiae, 1768.

Mackay, A, 'Crystallography and the Penrose pattern', *Physica* A (1981)

Penrose, R, 'Pentaplexity: a class of nonperiodic tilings of the plane', *Eureka* **39** (1978) (reprinted in *The Mathematical Intelligencer* **2** (1979) 32–7).

Steno, N, *De Solido intra Solidum Naturaliter Conteno*, Florence, 1669.

Zassenhaus, H, 'Über einen Algorithmus zur Bestimmung der Raumgruppen', *Commentaria Mathematica Helvetici* **21** (1948) 117–41.

General works

Any and all of the following can serve as a good introduction to the history of the subject:

Burckhardt, J J, *Die Symmetrie der Kristalle von René-Just Haüy sur kristallographischen Schule in Zürich*, Basel, Birkhäuser, 1988.

Burke, J, *Origins of the Science of Crystals*, Berkeley and Los Angeles, University of California Press, 1966.

Burckhardt, J J, *Die Symmetrie der Kristalle*, Basel, Birkhäuser, 1988.

Ewald, P, editor, *Fifty Years of X-ray Diffraction*, International Union of Crystallography, Oosthoek, Utrecht, 1962.

Metzger, H, *La Genèse de la Science des Cristaux*, Paris, 1922. Reprinted, 1966, A Blanchard.

Shafranosvki, I I, *History of Crystallography: from ancient times to the beginning of the* 19th *century*, Leningrad, Nauka, 1978 (in Russian).

Symmetry

There are many interesting books on symmetry, a vast subject which is at once aesthetic, philosophical, and technical. Here are a few of them.

Coxeter, H S M, Emmer M, Penrose R and Teuber M L, *M. C. Escher: Art and Science*, Amsterdam, North-Holland, 1986. (Proceedings of the International Congress on M C Escher in Rome, Italy, 26–28 March, 1985.)

Fejes-Tóth, L, *Regular Figures*, New York, Pergamon, 1964. (A fascinating exploration of the geometry of regular figures, including Delone's construction of Fedorov's parallelohedra.)

Gombrich, E H, *The Sense of Order: a study in the psychology of decorative art*, Ithaca, Cornell University Press, 1979. (A fascinating discussion of ornamental patterns, by one of today's leading art historians.)

Hargittai, I, editor, *Symmetry: Unifying human understanding*, Oxford, Pergamon Press, 1986. (A collection of 65 articles on symmetry in a wide variety of fields.)

MacGillavry, C, *Fantasy and Symmetry*, New York, H Abrams, 1976. (Formerly titled *Symmetry Aspects of M.C. Escher's Periodic Drawings*.)

Schattschneider, D, *Visions of Symmetry: the notebooks, periodic drawings, and related works of M. C. Escher*, New York, W H Freeman, 1990. (An interpretation of Escher's own methods, based on his unpublished notebooks.)

Senechal, M and Fleck, G, *Patterns of Symmetry*, Amherst, University of Massachusetts Press, 1977

Weyl, H, *Symmetry*, Princeton, Princeton University Press, 1952. (This is the great classic, whose influence can be seen in most subsequent work on the subject.)

Books on Mathematical Crystallography

The following books include much material relevant to the topics discussed in this one.

Brown, H, Bülow, R, Neubuser, J, Wondratschek, H and Zassen-haus, H, *Crystallographic Groups of Four-Dimensional Space*, New York, John Wiley and Sons, 1978. (The complete tables of the four-dimensional space groups.)

Buerger, M, *Elementary Crystallography*, New York, Wiley, 1956. (A thorough discussion of crystal symmetry, by one of this century's leading crystallographers.)

Burckhardt, J J, *Dic Bewegungsgruppen der Kristallographie*, Basel, Birkhäuser, 1946; 2nd edition, 1966. (An exposition of space group theory from the modern group-theoretic point of view.)

Conway, J H and Sloane, N J A, *Sphere Packings, Lattices, and Groups*, Springer Verlag, New York, 1988. (An encyclopedic tome on the interrelations among these three topics.)

Coxeter, H S M, *Regular Polytopes*, Dover, New York, 1973. (The classic work on regular figures in any dimension.)

Engel, P, *Geometric Crystallography*, Riedel, Dordrecht, 1986. (A sophisticated account of the geometric aspects of mathematical crystallography.)

Engel, P, Michel, L, and Senechal, M, *Lattice Geometry*, in preparation.

Galiulin, R G, *Crystallographic Geometry*, Moscow, Nauka, 1984 (in Russian). (An elegant geometrical formulation of crystallographic concepts, including the theory of (r, R) systems.)

Grünbaum, B and Shephard, G S, *Tilings and Patterns*, W H Freeman, New York, 1987. (The definitive but very accessible work on all aspects of tilings and patterns, including colored patterns, nonperiodic patterns, and many unsolved problems.)

Hahn, T, editor, *International Tables for X-Ray Crystallography*, Vol. A, Space-group symmetry, Dordrecht, Reidel, 2nd edition, 1988.

Hahn, T and Wondratschek, H, *Symmetry of Crystals: introduction to International Tables for Crystallography*, Vol. A, 2nd revised edition, 1989, preprint.

Loeb, A L, *Space Structures: their harmony and counterpoint*, Reading, Addison Wesley, 1976. (The emphasis on this readable book is on the combinatorial properties of networks.)

Schwarzenberger, R L E, *N-dimensional Crystallography*, London, Pittman, 1980. (A pioneering monograph.)

Selected articles

Michel, L and Mozryzmas, J, 'Les concepts fondamentaux de la cristallographie', *C. R. Acad. Sci. Paris*, **308**, Sér. II (1989) 151–8.

Schattschneider, D, 'The plane symmetry groups: their recognition and notation', *American Mathematical Monthly* **85** (1978) 439–50.

Senechal, M, 'The algebraic Escher', *Structural Topology* **15** (1988) 31–42.

Senechal, M, 'Which tetrahedra fill space?', *Mathematics Magazine* **54** (1981) 227–43.

Senechal, M, 'A simple characterization of the subgroups of space groups', *Acta Crystallographica* A **36** (1980) 845–50.

Senechal, M, 'Finding the finite subgroups of $O(3)$', *American Mathematical Monthly* **97** (1990) 329–35.

Senechal, M, 'Color Symmetry', *Computers and Mathematics with Applications* **16** (1988) 545–53.

Senechal, M and Taylor, J, 'Quasicrystals: the view from Les Houches', *The Mathematical Intelligencer* **12** (1990) 54–64.

Other books and articles quoted or referred to in the text

Barlow, W and Miers, H, 'The Structure of Crystals, I.' in *Report of the Meeting of the British Association for the Advancement of Science*, 71st meeting, London, 1901, 297–337.

Donnay, J D H and Harker, D, 'A new law of crystal morphology extending the law of Bravais', *Am. Mineral.* **22** (1937) 446–67.

Engel, P, 'Ueber Wirkungsbereichsteilungen mit kubischer Symmetrie', *Z. Kristallogr.* **157** (1981) 259–75.

Frost, R, 'Mending Wall'.

Needham, J, *Order and Life*, New Haven, Yale University Press, 1936. (Paperback edition, MIT Press.)

Rogers, C A, *Packing and Covering*, Cambridge, Cambridge University Press, 1964.

Shechtman, D, Bleck, I, Gratias, D and Cahn, J, 'Metallic phase with long-range orientational order and no translational symmetry', *Physical Review Letters* **53** (1984) 1951–4.

Shubnikov, A V, 'Symmetry and Antisymmetry of Finite Figures', Moscow, 1951.

Willson, S, 'Limiting spheres for configurations', *Journal of Computer and System Sciences* **15** (1977) 243–61.

Wulff, G, 'Zur Theorie der Kristallhabitus', *Z. Kristallogr.* **45** (1908) 433–73.

Wolff, P M de, *Acta Crystallographica* A **30** (1974) 777–85.

Index